2068 Scarlet Tiger *Callimorpha dominula 50mm*

This spectacular day-flying moth flew into the kitchen through an open door and laid eggs on the cooker!

2033 Black Arches female 50mm on a wet morning

# ACKNOWLEDGEMENTS

*These records would not be available at all were it not for the kindness of Giles Sturdy and his family who allowed me to run my mercury vapour (m.v.) lamps all over Trigon. It has been a privilege to be allowed access to all parts of this fascinatingly varied estate. I must particularly thank Robert Sturdy for his enthusiastic help with habitat management on the Speckled Footman site, including allowing me to burn off part of it, even though this did get slightly out of hand! The efforts are bearing fruit.*

*My sincere thanks also to Dr Phil Sterling, county micromoth recorder, for his endless patience in identifying hundreds of tiny specimens and photographs in spite of my continuing ignorance. An afternoon spent inspecting tree trunks with Phil will produce more micros than I would believe possible and all identified without further reference. His knowledge is astonishing and enthusiasm boundless.*

*I also owe a huge debt of gratitude to Peter Davey, the county macromoth recorder. He has continually encouraged me and shown me how to search out new habitats for target species. He has the miraculous ability to identify moths at a glance even in flight. My highest total of 125 macros recorded in one night was with Peter's assistance. Moths that I might have overlooked or mis-identified never seem to faze him, even the pugs!*

*Others have helped with advice, hints and specimens for photography (Phil, Peter, Dave Foot, Neil Sherman et al), for which many thanks. I am particularly indebted to Mark Parsons at Butterfly Conservation and Phil Sterling (again!) for so kindly proof-reading the draft manuscript and correcting the many errors.*

*Finally, my most heartfelt thanks to my wife, Eve, to whom this is dedicated. She has put up with being a 'moth widow' without complaint, and seems to understand the need for all my night-time and early morning excursions. She never complains about the necessity for a very bright light in the garden on summer nights, nor of finding stacks of pots in the fridge with livestock (i.e. moths) waiting to be photographed!*

*Chris Manley*

*January 2006*

# FOREWORDS

*This collection of beautiful photographs has opened my eyes to a new aspect of the natural world. I had no idea there were so many different species of moths on Trigon, each one with its own fragile habitat. This collection reinforces the sense of responsibility and the privilege that comes with owning and managing an estate like Trigon.*
*Many thanks Chris for such a fine contribution to our understanding.*

G. P. Sturdy
Trigon House

*****

*I first met Chris in the mid 1990s at a public demonstration by the light of my blue moth lamp on an aptly named Dorset heathland, Slop Bog. His fascination for moths, and his attention to detail was head and shoulders above others present that night. He has never let up and this book is a personal manifestation of Chris's wonder for a part of our natural world. I am delighted that today's technology has enabled him to turn an amateur enthusiasm into a professional publication and a work of art that will surely enhance our enjoyment of Dorset's unique environment.*

Dr. Phil Sterling, F.R.E.S.
Natural Environment Manager
Dorset County Council

*****

1913 Canary-shouldered Thorn 45mm on convolvulus

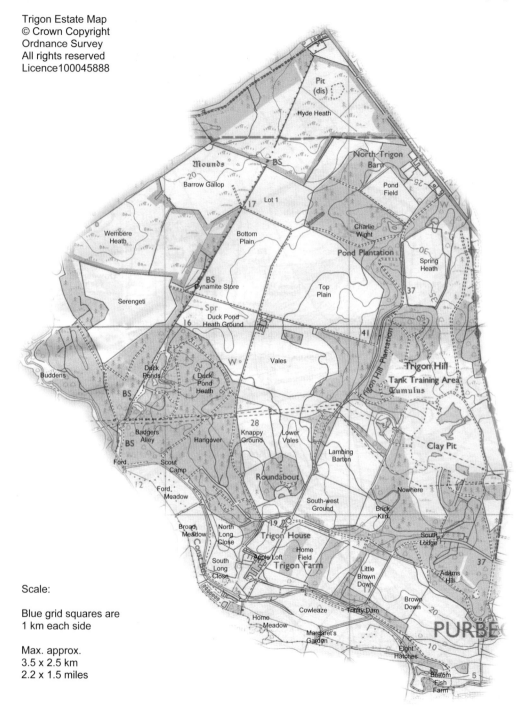

Scale:

Blue grid squares are
1 km each side

Max. approx.
3.5 x 2.5 km
2.2 x 1.5 miles

# TRIGON ESTATE

*This private estate of some 1500 acres has been in the hands of the Sturdy family since the middle of the nineteenth century when Queen Victoria reigned.*

*The estate was originally bought as a sporting estate suitable for hunting, shooting and fishing as a country retreat from city life in London.*

*The quality of land for farming was not the prime consideration so there are many areas of prime diverse habitat such as heathland, bog, old oaks, mature woodland and young plantations. There are ponds, streams, the River Piddle, damp ditches, thick hedges, copses and spinneys; also clay and gravel pits.*

*This fascinating variety of habitats is supplemented by having Poole Harbour to the east, the Purbeck hills and coast to the south and chalk downland to the north, often providing interesting records of moths not actually resident on the estate but migrating on the breeze. Some migrants like the tiny Diamond-back Moth Plutella xylostella come from all over Europe and even North Africa.*

*Continuity of ownership and careful management have created havens for some of Britain's rarest moths (Lepidoptera : scale-winged insects) such as Speckled Footman on the front cover and Blair's Wainscot, rediscovered in 1997 having been thought extinct since 1952.*

*****

# An introduction to the world of moths.

*Moths as a wildlife subject are thrilling, exciting and rewarding to study.*

*You can find them almost everywhere, from underwater (well, the larvae at any rate! Moths cannot actually fly underwater) like Brown China Mark, to mountain tops, in country or town, gardens, fields, woods, beaches. Even indoors where they may not be wanted, such as clothes moths, or roosting in sheds through the winter like the stunning Herald.*

*There are many, many different moths, around 2,500 recorded in this country. Lots are very small and sometimes difficult to identify, but there are about 900 macro (larger) species to be found. Almost any school or garden should be able to build a list of several hundred species without much difficulty. The Trigon list of 705 is illustrated with photographs where I have them, and supplemented with additional pictures of moths not yet seen on the estate.*

*They are amazingly beautiful and varied and you never know what might turn up. All moths have a recognised flight period so what you see will depend on the season as well as weather and habitat.*
*Some are also double-brooded so may be on the wing for a couple of months early in the year and again later on like the beautiful Gold Spot.*
*Some seem to be continuously brooded and can appear at almost any time. The Australian import Epiphyas postvittana sits on my kitchen window on any mild night of the year. (We have no blinds in the kitchen and attract all sorts of moths that way!)*
*Others have a ridiculously short flight time like Blair's Wainscot which seems only to fly from 7.15 to 7.30 at dusk in the first week of October. It then seems to fly again at about 9pm when it may come to light. This short period possibly explains how it was thought extinct for 45 years after its only known marshy habitat on the Isle of Wight was drained and burnt in 1952, before being discovered at a new site in Dorset in 1997.*
*There are moths on the wing even in mid-winter, such as Winter Moth or December Moth, and I've seen Oak Beauty by the porch light on New Year's Eve.*

*In years of warm southerly winds blowing up from continental Europe all sorts of migrants can appear.*
*Most years Silver Y will be seen in good numbers, with some of the early arrivals laying eggs and producing a later brood. Sometimes the huge, pink-bodied Convolvulus Hawk can be seen. They have a passion for tobacco plants, hovering in front of the flowers to sip nectar, like a flying mouse.*
*Occasionally also there can be very rare species like the pretty little pyrale Evergestis limbata, and 2005 produced great excitement when two Oak Rustic appeared at Durlston. These were the first ever seen in Dorset and the first ever in mainland Britain was only the night before in Hampshire! They feed on evergreen oak, Quercus ilex, so perhaps they might become established as this is a common tree in this part of Dorset.*

# First catch your moth..........

*Moths are not difficult to find; everyone knows they are attracted to lights and will probably have seen an occasional one at a lighted window in summer. As with all generalisations there are plenty of exceptions. Although most do fly at night some fly only in sunshine like Speckled Yellow, and many of the micros take no notice at all of lights. There are other ways of finding them.*

*The aim is to attract lots of moths! This is not difficult with a modest investment in the right sort of lamp ( see Reading & References at the end of the book).*
*Insects can see the 'colour' ultra (beyond) violet, which we cannot, and are attracted to it (evening primrose flowers have ultra violet markings on their petals to guide bees to the centre of the flower, though to us they just look yellow). No one knows why moths come to lights but we can harmlessly exploit their attraction by using a mercury vapour (m.v.) lamp which emits plenty of ultra violet. The 40 watt blue light of fly traps in food shops will attract moths if the bulb is used on its own without the fly-killing part; good in urban gardens. Most moth traps use a brighter bulb of either 80 or 125 watts, but these may, with good reason, be unpopular with neighbours if not prevented from shining into their windows!*

Having bought a mercury vapour lamp and the necessary 'choke' or 'ballast' to go with it and some cable to reach from a convenient socket to an outdoor site you are virtually ready to start attracting moths.

Moths will come to the light anyway, so it is not essential to have any sort of trap. It can be quite a social event having the light on a white sheet on the lawn with everyone inspecting new arrivals as they appear.

A trap, though, channels the natural instincts of moths to the observer's advantage. When moths approach the light they will circle and then dive down below it. If the light is over a box of some sort and containing a few egg trays for the creatures to rest on they will sit quietly(ish!) on the darker sides of the trays. These can be lifted out, inspected, handed round and replaced without disturbing the moths.

Any sort of container will do; I used a cardboard box for a couple of years with a sheet of glass a metre above it to keep the rain off. Expensive traps are, of course, available and very well designed to maximise the catch, but in my experience most people make their own. The light is the important part, not the design of trap.

Siting and weather conditions can heavily influence the catch. Clear, cold nights and windy conditions are unfavourable. The ideal has to be a warm, humid, cloudy calm evening, and if you get these conditions in mid-summer you will be swamped with 500 plus (sometimes many more) moths all competing for space round your lamp! The calm before a huge thunderstorm will produce the most exciting mothing ever, but you have to be prepared to get absolutely soaked as you pack up just too late in the pouring rain!

I prefer to inspect my traps in the morning. It's cool so the moths are quiet and there is natural light if there is something to photograph. There are also far fewer unwanted mosquito-type insects flying about which can make trapping uncomfortable at night. Once the catch has been inspected, identified and recorded moths can be tapped off unharmed into nearby bushes and vegetation.

Apart from the practical advantages, early dawns have a magic of their own and you have the world to yourself in peace and quiet.

## *Micros.*

There are around twice as many micros as macros and the split between the two is by no means clear.

When my wife finally invested in Bernard Skinner's *Macrolepidoptera of the British Isles* for me I was disappointed to find that one of the first moths I attempted to identify was not shown. When I contacted Mr Skinner he kindly explained that he had not made an error and that the Small Magpie, although a fair-sized moth, is actually a pyrale, one of the micros! He suggested I get Barry Goater's *British Pyralid Moths*, so that was another investment that had to be made!

Although some micros, especially the pyrales, will come to light, there are many more that will not, or will only come to weak light like the kitchen window and not a bright m.v. lamp.

Although many of the micros do live up to their name and are almost too small to see ( *Phyllonorycter leucographella* resting is less than 4mm long and the thickness of a fine pencil mark) there are tell-tale signs for many. If you have ever wondered about the white squiggly lines on the leaves of various plants these are made by moth larvae. Known as leaf-miners, the type of leaf and style of mine can provide an identification. *P. leucographella* above is a recent arrival in Britain, probably from the nursery trade, that lives inside the leaves of *Pyracantha* (firethorn) bushes, creating a 'blister' down the centre of each leaf as the larva feeds under the surface layer. It is abundant on the Pyracantha by my front door but even when the adults are flying they are hard to see.

Just looking close up in long grass or carefully inspecting lichen-covered bark on a tree in summer is likely to produce a surprising variety of species.
The picture of *Dichrorampha petiverella* was taken in a small patch of rough grass that I'd not mown at the bottom of the garden. *Chrysoteuchia culmella* or Garden Grass-veneer is a pyrale that can be abundant in rough grass almost anywhere and is easily disturbed in sunlight, showing up well in flight but not so easily seen resting head down on a grass stem.

# PHOTOGRAPHY

*The photographs each have a caption in three parts.*
*First is the taxonomic number taken from the Bradley 2000 Lepidoptera Checklist.*
*Next is the name. This is either in Latin (the scientific name so that the same moth is correctly identified in any language) for micros or English for macros, as most micros do not have vernacular names in regular use. As the names suggest, micros are generally smaller than macros but there are plenty of confusing exceptions with big micros and small macros! Where given f. is female, m. is male.*
*Lastly the approximate wingspan in millimetres is given. Usually a larger moth merits a larger picture but they are not to scale. A big photo can just mean that I am pleased with it! More information can be found for each moth pictured in the log order list later on. The pictures themselves are arranged as nearly as possible in their taxonomic order, meaning that similar moths can easily be compared, though differences can sometimes be hard to spot!*

*Photography is a way of responsible 'collecting' and as such can provide important records easily shared but my main incentive is that moths are astonishingly beautiful. Most of these pictures were taken with a digital camera, though there are just a few included from film days. (The one below is scanned from a print but is a beautiful example of a moth hiding in plain sight). Some I am very pleased with, others I would hope to improve on. Many of these moths have been attracted to light and then been persuaded to pose nicely. In the cool of a summer dawn this can be relatively easy, but the alternative is to take the moth home and keep it in the fridge for a while. In theory this will calm the moth but it can still sometimes take an hour to get the shot.*
*One extremely calm moth was an Oak Beauty in February 2006 that sat through a snowstorm on the wall by my front door, and then several nights at −5°C before the weather improved enough for it to fly again. I had a whole week to take a picture!*

2247 Merveille du Jour 45mm camouflaged on lichen

14 Ghost Moth f. 55mm

17 Common Swift 30mm

150 Adela reaumurella f. 15mm

286 Caloptilia alchimiella 12mm

148 Nemophora degeerella m. 20mm

130 15mm
Incurvaria
masculella m.

544 14mm
Coleophora
albicosta

282 Caloptilia elongella 15mm

296 Calybites phasianipennella
form. quadruplella 10mm

173 Festoon m. 21mm

15 Orange Swift m. upper 33mm, f. lower 46mm

385
Anthophila
fabriciana
12mm

161 Leopard m. 50mm

373 Currant Clearwing 19mm

424
Yponomeuta
evonymella 20mm

452 Ypsolopha nemorella 23mm

332a 7mm
Phyllonorycter leucographella

287 Caloptilia robustella 12mm

*Plate* I

458 Ypsolopha alpella 16mm

460 Ypsolopha parenthesella 18mm

461 Ypsolopha ustella 17mm

462 Ypsolopha sequella 19mm

654 Pleurota bicostella 21mm

288 Caloptilia stigmatella 12mm

663 Diurnea fagella  25mm

695 Agonopterix alstromeriana 18mm

854 Anacampsis blattariella 16mm

874 Blastobasis decolorella 20mm

936 Cochylimorpha straminea 13mm

938 Agapeta zoegana 20mm

966 Cochylis atricapitana 14mm

977 Archips podana m. 18mm

977 Archips podana f. 25mm

993 Clepsis spectrana 20mm

998 Epiphyas postvittana 20mm

1006 Epagoge grotiana 15mm

976 Archips oporana 28mm

1061 Acleris literana 18mm

656 Tachystola acroxantha 14mm

1048 Acleris variegana 15mm

1120  14mm Ancylis mitterbacheriana

714  20mm Agonopteryx yeatiana

1063 Celypha striana 20mm

1082 Hedya pruniana 17mm

1045 Acleris notana 17mm

1025 Tortricodes alternella 21mm

1033 Tortrix viridana 22mm

1048 Acleris variegana 15mm

1092 Apotomis turbidana 20mm

1111 Bactra lancealana 20mm

*Plate 2*

1133 Epinotia bilunana 15mm

1134 Epinotia ramella 15mm

1134 Epinotia ramella 15mm

1138 Epinotia nisella 15mm

1201 Eucosma cana 20mm

1184 Epiblema scutulana 20mm

1204 Thiodia citrana 15mm

1211 Rhyacionia pinicolana 20mm

1273 Dichrorampha petiverella 12mm

1288 Alucita hexadactyla 15mm

1292 Calamotropha paludella f. 30mm

1174 Epiblema cynosbatella 20mm

1212 Rhyacionia pinivorana 17mm

1309 Agriphila geniculea 21mm

1299 Crambus hamella 25mm

1316 Catoptria falsella 20mm

1329 Donacaula forficella m. 30mm

1333 Scoparia pyralella 20mm

1307 Agriphila latistria 25mm

1302 Crambus perlella form.warringtonellus 25mm

1302 Crambus perlella 25mm

1303 Agriphila selasella 25mm

1305 Agriphila tristella 27mm

*Plate 3*

1385 Ebulea crocealis 24mm

1348 Parapoynx stratiotata f. 30mm

1380 Phlyctaenia perlucidalis 22mm

1350 Nymphula nitidulata 20mm

1356.1 Evergestis limbata 23mm

1378 Phlyctaenia coronata 24mm

1413 Hypsopygia costaliis 20mm

1376 Eurrhypara hortulata 32mm

1362 Pyrausta purpuralis 20mm

1345 Elophila nymphaeata 28mm

1377 Perinephela lancealis m. 34mm

1383 Psammotis pulveralis 25mm

1382 Anania verbascalis 23mm

1389 Udea fulvalis 26mm

1395 Udea ferrugalis 21mm

1375 Ostrinia nubilalis m. 34mm

1390 Udea prunalis 23mm

1398 Nomophila noctuella 30mm

1424 Endotricha flammealis 20mm

1405 Pleuroptya ruralis 35mm

*Plate* 4

1425 Galleria mellonella f. 30mm

1428 Aphomia sociella f. 32mm

1452 Phycita roborella 27mm

1458 Myelois circumvoluta 34mm

1474 Ephestia parasitella 16mm

1441 Oncocera semirubella 28mm

1454.2 Dioryctria syvestrella 32mm

1470 Euzophera pinguis 26mm

1458 Myelois circumvoluta variant. 34mm

1442 Pempelia palumbella 25mm

1485 Phycitodes maritima 20mm

1488 Agdistis bennetii 28mm

1494 Capperia britanniodactyla 20mm

1524 Emmelina monodactyla 23mm

1517 Adaina microdactyla 15mm

1495 Marasmarcha lunaedactyla 20mm

1497 Amblyptilia acanthadactyla 20mm

1507 Stenoptilia zophodactylus 20mm

1513 Pterophorus pentadactyla 33mm

Plate 5

1643 Emperors mating
f. upper 70mm
m. lower 55mm

1631 December Moth m. 31mm

1634 Lackey m. 28mm

1637 Oak Eggar f. 75mm

1645 Scalloped Hook-tip 28mm

1646 Oak Hook-tip 30mm

1638 Fox Moth f. left 65mm m. right 50mm

*Plate 6*

1647 Barred Hook-tip 29mm

1648 Pebble Hook-tip 32mm

1652 Peach Blossom 35mm

1654 Figure of Eighty 37mm

1651 Chinese Character 20mm

1657 Common Lutestring 30mm

1659 Yellow Horned 38mm

1663 March Moth 23mm

1660 Frosted Green 35mm

1653 Buff Arches 40mm

1666 Large Emerald 45mm

1669 Common Emerald 27mm

1670 Small Grass Emerald 21mm

*Plate* 7

1673 Small Emerald  30mm

1675 Dingy Mocha 25mm

1677 Birch Mocha 23mm

1680 Maiden's Blush 28mm

1681 Clay Triple-lines 28mm

1682 Blood-vein 38mm

1689 Mullein Wave 24mm

1690 Small Blood-vein 28mm

1698 Purple-bordered Gold 15mm

1701 Dotted-border Wave 22mm

1713 Riband Wave 32mm

1716 Vestal  25mm

1716 Vestal  25mm

1705 Dwarf Cream Wave 20mm

1708 Single-dotted Wave 20mm

1715 Plain Wave 30mm

1716 Vestal 25mm

1719 Oblique Carpet 25mm

*Plate 8*

1720
Gem m.
13mm

1720
Gem f.
20mm

1722
Flame
Carpet
24mm

1724 Red Twin-spot Carpet 23mm

1728
Garden
Carpet
25mm

1732 Shaded Broad-bar 30mm

1734 July Belle 25mm

1742 Yellow Shell  26mm

1746 Shoulder Stripe  26mm

1747 Streamer 25mm

1793 Cloaked Carpet  32mm

1752 Purple Bar 25mm

1754 Phoenix 35mm

1755 Chevron 32mm

1760 Red-green Carpet 27mm

*Plate  9*

1764 Common Marbled Carpet 30mm

1784 Common Marbled Carpet 30mm

1765 Barred Yellow 25mm

1766 Blue-bordered Carpet 25mm

1767 Pine Carpet 25mm

1768 Grey Pine Carpet 25mm

1769 Spruce Carpet 25mm

1771.1 Cypress Carpet 25mm

1773 Broken-barred Carpet 28mm

1776 Green Carpet 24mm

1777 July Highflyer 30mm

1778 May Highflyer 27mm

1779 Ruddy Highflyer 27mm

1789 Scallop Shell 30mm

1792 Dark Umber 30mm

1794 Sharp-angled Carpet 26mm

1795 November Moth 35mm

1797 Autumnal 40mm

1799 Winter Moth m. 25mm

1800 Northern Winter m. 32mm

*Plate* 10

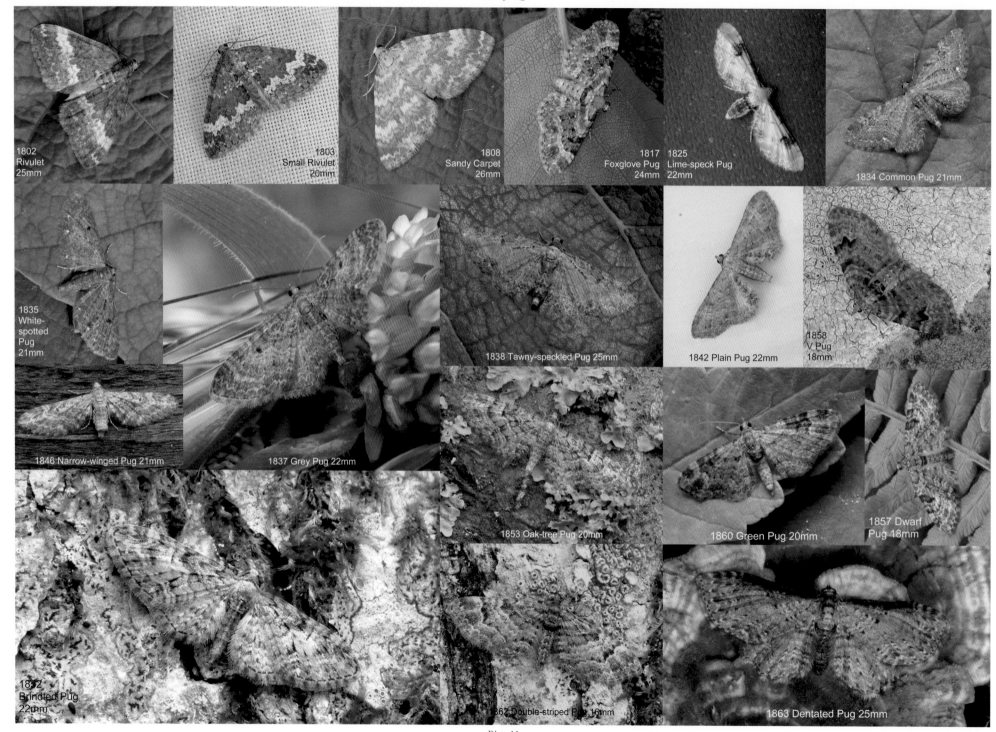

1802
Rivulet
25mm

1803
Small Rivulet
20mm

1808
Sandy Carpet
26mm

1817
Foxglove Pug
24mm

1825
Lime-speck Pug
22mm

1834 Common Pug 21mm

1835
White-
spotted
Pug
21mm

1838 Tawny-speckled Pug 25mm

1842 Plain Pug 22mm

1858
V Pug
18mm

1846 Narrow-winged Pug 21mm

1837 Grey Pug 22mm

1853 Oak-tree Pug 20mm

1860 Green Pug 20mm

1857 Dwarf
Pug 18mm

1852
Brindled Pug
22mm

1862 Double-striped Pug 16mm

1863 Dentated Pug 25mm

*Plate* 11

1864 Streak 34mm

1867 Treble Bar 40mm

1874 Dingy Shell 20mm

1875 Small Yellow Wave 20mm

1879 Seraphim 28mm

1881 Early Tooth-striped 30mm

1881 Early Tooth-striped aberration

1883 Yellow-barred Brindle 25mm

1888 Scorched Carpet 25mm

1884 Magpie 40mm

1887 Clouded Border 25mm

1889 Peacock 30mm

1893 Tawny-barred Angle 30mm

1894 Latticed Heath 26mm

1901 Little Thorn 25mm

*Plate* 12

1902 Brown Silver-line 30mm

1904 Scorched Wing 30mm

1905 Horse Chestnut 25mm

1906 Brimstone 35mm

1907 Bordered Beauty 25mm

1909 Speckled Yellow 26mm

1913 Canary-shouldered Thorn 40mm

1914 Dusky Thorn 44mm

1917 Early Thorn 40mm

1919 Purple Thorn 44mm summer

1919 Purple Thorn 35mm spring

1920 Scalloped Hazel 44mm

*Plate* 13

1922 Swallow-tailed Moth 55mm

1923 Feathered Thorn 44mm

1923 Feathered Thorn 44mm

1923 Feathered Thorn 44mm

1926 Pale Brindled Beauty 40mm

1927 Brindled Beauty 44mm

1930 Oak Beauty 50mm

1932 Spring Usher 34mm

1932 Spring Usher 34mm

*Plate* 14

1931 Peppered f. 60mm

1931 Peppered m. usual & *carbonaria* form. 50mm

1933 Scarce Umber m. 35mm

1934 Dotted Border m. 40mm

1935 Mottled Umber m. 40mm

1936 Waved Umber 40mm

1938 Bordered Grey 40mm

*Plate* 15

1937 Willow Beauty 40mm

1941 Mottled Beauty 45mm

1941 Mottled Beauty 45mm

1944 Pale Oak Beauty 45mm

1945 Brussels Lace 30mm

1947 Engrailed 40mm

1949 Square Spot 40mm

1955 Common White Wave 30mm

1954 Bordered White m. 35mm

1958 Clouded Silver 28mm

1960 Early Moth 30mm

1962 Barred Red 40mm

*Plate* 16

1964 Annulet 30mm

1968
Yellow
Belle m.
28mm

1970 Grass Wave m. 35mm

1992 Small Elephant Hawk-moth 45mm

1976 Privet Hawk-moth 110mm

*Plate* 17

1972 Convolvulus Hawk-moth 110mm

1978 Pine Hawk-moth 90mm

1979 Lime Hawk-moth 65mm

1980 Eyed Hawk-moth 80mm

*Plate* 18

1981 Poplar Hawk-moth m. usual  80mm

1981 Poplar Hawk-moth f. buff form 90mm

1991 Elephant Hawk-moth 70mm

1992 Small Elephant Hawk-moth 45mm

*Plate* 19

2006 Lesser Swallow Prominent 45mm

1994 Buff-tip 50mm

2003 Pebble Prominent 40mm

1997 Sallow Kitten 35mm

2000 Iron Prominent 45mm

1999 Lobster Moth 60mm

1995 Puss Moth 70mm

2005 Great Prominent 60mm

*Plate* 20

2010
Scarce
Prominent
40mm

2011
Pale
Prominent
45mm

2008 Coxcomb Prominent 40mm

2014 Marbled Brown 40mm

2015 Lunar Marbled Brown 35mm

2017 Small Chocolate Tip 25mm

2019 Chocolate Tip 35mm

2026 Vapourer m. 30mm

2027 Dark Tussock m. 45mm

2033 Black Arches m. 40mm

*Plate* 21

2028
Pale
Tussock m.
45mm

2028 Pale Tussock f. 60mm

2031 White Satin 45mm

2040 Four-dotted Footman 32mm

2035
Round-winged Muslin 20mm

2043
Orange
Footman
25mm

2039
Red-necked
Footman
25mm

2049 Buff Footman f. 35mm

2053 Speckled Footman m. 30mm

2059 Clouded Buff m. 40mm

2044 Dingy
Footman 32mm

2037 Rosy Footman 25mm

2047 Scarce Footman 30mm

2053 Speckled Footman f. 30mm

2058 Cream-spot Tiger 60mm

*Plate* 22

2057 Garden Tiger 70mm

2060 White Ermine 40mm

2061 Buff Ermine 40mm

2063 Muslin m. 35mm

2063 Muslin f. 35mm

2069 Cinnabar 40mm

2068 Scarlet Tiger 50mm

*Plate* 23

2076 Kent Black Arches 24mm

2078 Least Black Arches 18mm

2081 White-line Dart 35mm

2085 Archer's Dart 35mm

2090 Crescent Dart 40mm

2091 Dark Sword-grass 45mm

2092 Shuttle Shaped Dart 30mm

2098 Flame 35mm

2102 Flame Shoulder 30mm

2087 Turnip 40mm

2107 Large Yellow Underwing 55mm

2110 Broad-bordered Yellow Underwing 60mm

2117 Autumnal Rustic 35mm

*Plate* 24

2119
Pearly
Underwing
45mm

2120 Ingrailed Clay 35mm

2122 Purple Clay 35mm

2109 Lesser Yellow Underwing 45mm

2126 Setaceous Hebrew Character 35mm

2128 Double Square-spot 44mm

2132 Neglected Rustic 40mm

2133 Six-striped Rustic 35mm

2134 Square-spot Rustic 35mm

2135 Heath Rustic 32mm

*Plate* 25

2138 Green Arches 45mm

2139 Red Chestnut 35mm

2142 Beautiful Yellow Underwing 22mm

2147 Shears 35mm

2154 Cabbage 45mm

2156 Beautiful Brocade 36mm

2157 Light Brocade 42mm

2150 Grey Arches 50mm

2158 Pale-shouldered Brocade 40mm

2159 Dog's Tooth 36mm

2163 Broom 40mm

2166 Campion 34mm

*Plate* 26

2173 Lychnis 35mm

2177 Hedge Rustic 40mm

2178 Feathered Gothic 44mm

2179 Pine Beauty 32mm

2182 Small Quaker 30mm

2186 Powdered Quaker 40mm

2187 Common Quaker 35mm

2188 Clouded Drab 40mm

2189 Twin-spotted Quaker 40mm

2190 Hebrew Character 34mm

2192 Brown-line Bright-eye 34mm

2194 White-point 34mm

*Plate* 27

2195 Delicate 40mm

2199 Common Wainscot 35mm

2200 Mathew's Wainscot 35mm

2202 L-album Wainscot 35mm

2197 Southern Wainscot 35mm

2205 Shoulder-striped Wainscot 35mm

2237 Grey Shoulder-knot 38mm

2214 Chamomile Shark 45mm

2235 Tawny Pinion 40mm

2236 Pale Pinion 40mm

2240 Blair's Shoulder-knot 40mm

2221 Mullein 45mm

*Plate* 28

2243 Early Grey 32mm

2247 Merveille du Jour 45mm

2248 Brindled Green 34mm

2251 Deep-brown Dart 40mm
2252 Black Rustic 45mm

2255 Feathered Ranunculus 35mm

2256 Satellite 40mm

2262 Brick 36mm

2245 Green-brindled Crescent 40mm

2260 Dotted Chestnut 35mm

2264 Yellow-line Quaker ab. *obsoleta* & usual 35mm

2263 Red-line Quaker 35mm

*Plate* 29

2267 Beaded Chestnut 35mm

2268 Suspected 30mm

2269 Centre-barred Sallow 32mm

2272 Barred Sallow 30mm

2270 Lunar Underwing 35mm

2270 Lunar Underwing 35mm

2270 Lunar Underwing 35mm

2273 Pink-barred Sallow 30mm

2281 Alder 35mm

2280 Miller 40mm

2274 Sallow 35mm

2274 Sallow 35mm

*Plate* 30

2284 Grey Dagger 45mm

2289 Knot Grass 35mm

2293 Marbled Beauty 25mm

2297 Copper Underwing 50mm

2299 Mouse Moth 25mm

2301 Bird's Wing 25mm

2302 Brown Rustic 35mm

2300 Old Lady 70mm

2303
Straw Underwing 40mm

*Plate* 31

2305 Small Angle Shades 35mm

2306 Angle Shades 50mm

2130 Dotted Clay 38mm

2318 Dun-bar 35mm

2319 Lunar-spotted Pinion 30mm

2319 Lunar-spotted Pinion 30mm

2326 Clouded-bordered Brindle 40mm

2326 Clouded-bordered Brindle, form *combusta*

2330 Dusky Brocade 40mm

2335 Slender Brindle 34mm

2336 Double Lobed 30mm

2339 Tawny Marbled Minor 26mm

*Plate* 32

2340 Middle-barred Minor 25mm

2342 Rosy Minor 25mm

2345 Small Dotted Buff m. 25mm

2350 Small Wainscot 24mm

2352 Dusky Sallow 33mm

Left 2358 Saltern Ear 34mm. right 2360 Ear 30mm

2361 Rosy Rustic 35mm

2364 Frosted Orange 38mm

2370 Twin-spotted Wainscot 25mm

2371 Brown-veined Wainscot 25mm

2375 Large Wainscot 50mm

2376 Blair's Wainscot 28mm

*Plate* 33

2385
Small Mottled Willow 25mm

2380 Treble Lines 35mm

2391 Silky Wainscot 25mm

2394 Anomalous m. 35mm

2412
Silver Hook
22mm

2418
Cream-bordered
Green Pea 20mm

2422
Green Silver-lines
35mm

2403 Bordered Straw 35mm

2423 Oak Nycteoline 22mm

2423 Oak Nycteoline 22mm

2434 Burnished Brass 40mm

2441 Silver Y 36mm

2425 Nut-tree Tussock 35mm

*Plate* 34

2439 Gold Spot 38mm

2449 Dark Spectacle 37mm

2450 Spectacle 34mm

2432 Ni 32mm

2442 Beautiful Golden Y 40mm

2469 Herald 45mm

2473 Beautiful Hook-tip 28mm

2492 Small Fan-foot 24mm

2470 Small Purple-barred 18mm

2452 Red Underwing 80mm

2484 Pinion-streaked Snout 20mm

2474 Straw Dot 20mm

2477 Snout 38mm

*Plate* 35

# FURTHER PHOTOGRAPHY

*The following pages contain photographs of moths not yet seen on Trigon but which have been recorded in Dorset so might turn up some night.*
*There are also six that have not been seen in Dorset but which you may be lucky enough to see elsewhere. Kentish Glory and Pretty Pinion from Scotland, Silurian and Silky Wave from Wales, Fisher's Estuarine is in Essex and Silver Cloud occurs in Gloucestershire.*
*I am fortunate to have access to some of the National Trust lands locally such as Studland Bay with its dunes, heath and inland Little Sea and also the RSPB reserve at Arne. These are some prime habitats providing conditions for many scarce as well as common species.*

*Durlston Country Park produces many migrants from the Continent on southerly breezes. In spite of the number of illustrations included this is by no means an exhaustive list of all the moths to be found in this area, but hopefully the ones here will encourage further study. There are many more pictures to be made of species I have already seen as well as the multitudes that I have yet to discover.*
*However, if the photographs shown kindle some interest and enthusiasm for these stunning creatures I shall feel justified in composing this work. It is an absorbing hobby and most landowners are pleased to have the records to aid the continuing welfare of moths.*

1973 Death's-head Hawk-moth 120mm

*Plate* 36

5 *Micropterix calthella* mating 7mm

140 *Nematopogon swammerdamella* 18mm

436 *Pseudoswammerdamia combinella* 15mm

455 *Ypsolopha scabrella* 22mm

652 *Alabonia geoffrella* 20mm

700 *Agonopterix pallorella* 20mm

706 *Agonopterix nervosa* 18mm

727.1 *Metzneria aprilella* 17mm

726 *Metzneria metzneriella* 17mm

1481 *Homoeosoma sinuella* 20mm

162 Goat f. 70mm

160 Reed Leopard m. 30mm

1633 Small Eggar m. 34mm

729 *Isophrictis striatella* 8mm Egg-laying on tansy

950 *Aethes francillana* 15mm

965 *Cochylis hybridella* 15mm

*Plate* 37

980 Archips xylosteana 20mm

1036 Acleris forsskaleana 17mm

1062 Acleris emargana 20mm

1272 Pammene aurana 12mm

1367 Pyrausta cingulata 16mm

1435 Conobathra tumidana 22mm

1397 Mecyna asinalis 27mm

1399 Dolicharthria punctalis 25mm

1408 Palpita vitrealis 30mm

1465 Nephopterix angustella 22mm

1461 Assara terebrella 22mm

1457 Hypochalcia ahenella 24mm

1519 Euleioptilus carphodactyla 20mm

1503 Platyptilia ochrodactyla 25mm

1642 Lappet 70mm

1523 Oidaematophorus lithodactyla 28mm

1504 Platyptilia pallidactyla 25mm

*Plate* 38

1636 Grass Eggar  f. 70mm

1655 Poplar Lutestring 40mm

1658 Oak Lutestring 34mm

1662 Light Orange Underwing 33mm

1667 Blotched Emerald  30mm

1691 Rosy Wave  24mm

1662 Light Orange Underwing 33mm

1644 Kentish Glory mating & eggs  f. 70mm,  m.60mm

1704 Silky Wave 20mm

1714 Portland Ribbon Wave 25mm

1739 Wood Carpet 29mm

1781 Small Waved Umber 32mm

*Plate* 39

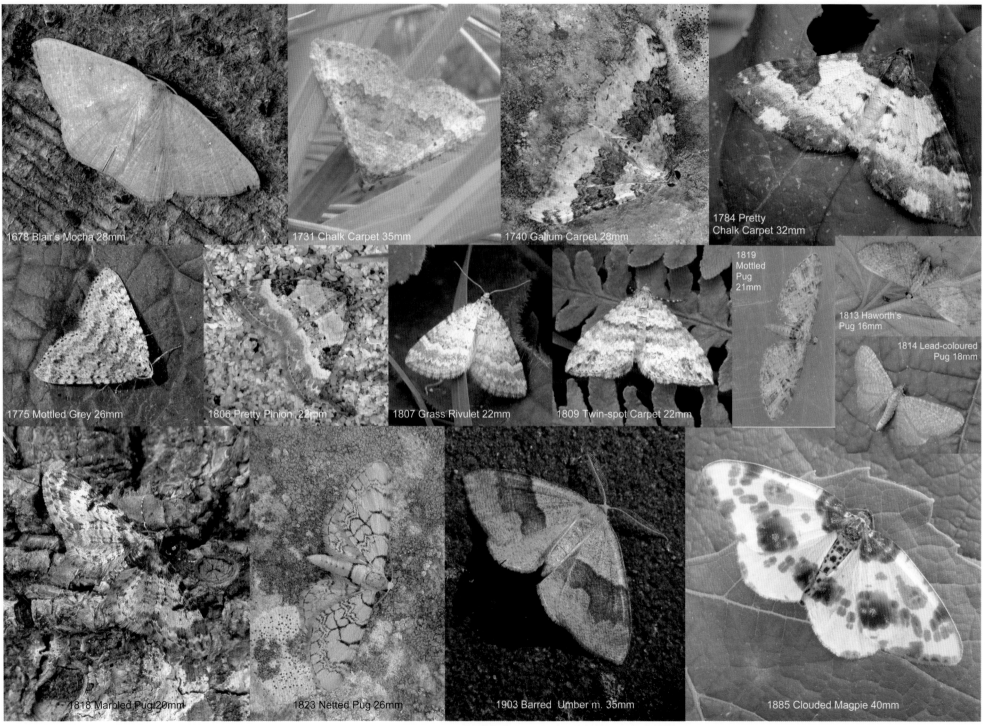

1678 Blair's Mocha 28mm

1731 Chalk Carpet 35mm

1740 Galium Carpet 28mm

1784 Pretty Chalk Carpet 32mm

1775 Mottled Grey 26mm

1806 Pretty Pinion 22mm

1807 Grass Rivulet 22mm

1809 Twin-spot Carpet 22mm

1819 Mottled Pug 21mm

1813 Haworth's Pug 16mm

1814 Lead-coloured Pug 18mm

1818 Marbled Pug 20mm

1823 Netted Pug 26mm

1903 Barred Umber m. 35mm

1885 Clouded Magpie 40mm

*Plate* 40

1957 White-pinion Spotted 27mm

1990 Striped Hawk-moth 80mm

2034 Gypsy m. 45mm

2075 Small Black Arches 22mm

2093 Sand Dart 34mm

2104 Northern Rustic 40mm

2108 Lunar Yellow Underwing  40mm

2148 Pale Shining Brown 50mm

2152 White Colon  37mm

2172 White Spot  34mm  on Nottingham catchfly

2175 Silurian m. 27mm

*Plate* 41

2176 Antler f. 35mm

2181 Silver Cloud 36mm

2191 Double Line 44mm

2209 Flame Wainscot 35mm

2201 Shore Wainscot 36mm

2226 Beautiful Gothic 30mm

2229 Brindled Ochre 50mm

2230 Feathered Brindle 38mm

2325 Crescent Striped 45mm

2252 Large Ranunculus 42mm

2264.1 Southern Chestnut 35mm

*Plate* 42

2246.1 Oak Rustic 30mm

2334 Rustic Shoulder-knot 40mm

2462 Mother Shipton 30mm

2463 Burnet Companion 30mm

2399 Bordered Sallow 34mm

2365 Fisher's Estuarine 50mm

2397 Small Yellow Underwing 18mm

2443 Plain Golden Y 40mm

*Plate* 43

## *Moth or Butterfly?*

Moths and butterflies are Lepidoptera on the same taxonomic list together.

Butterflies are a small group of only 59 species in Britain, generally distinguished from moths by having clubbed antennae and resting with wings folded over their body, as illustrated. They fly by day.

The life-cycle of all Lepidoptera is similar, as shown, generally spanning a year. Eggs are usually laid on the food plant, though some like the Swifts scatter eggs whilst flying.

The soft-bodied caterpillar eats a lot, changing its skin (moulting) about 5 times as it grows. These stages are called instars, and can last a few weeks or, rarely, as long as 5 years.

The pupa is formed, often underground, sometimes in a silk cocoon, from which the adult usually emerges the following year. The wings are soft initially but pump up and harden in an hour or so.

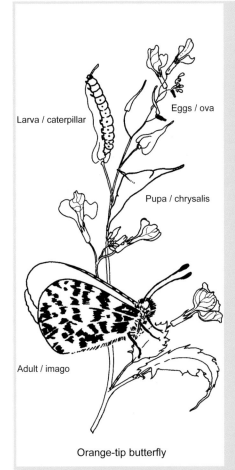

Larva / caterpillar

Eggs / ova

Pupa / chrysalis

Adult / imago

Orange-tip butterfly

## *Recording*

It is important to keep records of all moths seen, noting date, time, number, place (preferably with a grid reference) etc. Shorthand notebooks are good and over the years provide a fascinating history. Records should be submitted to your County Recorder at least annually who will incorporate them into the county database.

The Heritage Lottery Fund awarded £806,000 in 2006 to support the new National Moth Recording Scheme for four years.

See www.mothrecording.org.uk or www.butterfly-conservation.org

Butterfly Conservation is the "UK charity working towards a world where butterflies and moths can thrive for future generations to enjoy."

## *The Records* 1993 - 2005

This work started as a list of records from the Trigon estate, compiled because we were moving house. In the end we stayed put so I then developed the list somewhat!

Whilst this is a full list of everything that I have found there will be many more. The photographs are simply the ones that I already have as an enthusiastic amateur. I have also added further pictures of moths that I have not seen here, some of which may one day appear and some certainly will not! The Trigon list stands at 705 species but could perhaps reach 1,000. I need to work harder at identifying micros! There are 508 species illustrated in total.

The log number is taken from Bradley, J.D. 2000 Checklist of Lepidoptera recorded from the British Isles, the taxonomic order for all British Lepidoptera.

Latin names are shown for all moths, English names are shown for macros. Some larger or more common micros also have vernacular names but I have added micro, pyrale or plume as appropriate.

The relevant photographic plate number is shown where there is one.

The place and date of first sighting are noted. This only indicates that I have seen the moth at least once, without distinguishing between abundant species and rarities. The date can be an important aid to identification as most species have definite flight periods, e.g. the Spring Usher cannot be found in autumn!

All Trigon places start with T. and can be found on the map at the beginning. Other references can be either villages in Dorset or further afield.

The next two columns show national status and Dorset status if different.

| | |
|---|---|
| RDB 1,2 ,3 | Red Data Book. Top three national rarities, 1 is rarest. |
| pRDB | Provisional Red Data Book, scarce but insufficiently known. |
| BAP | UK Biodiversity Action Plan for targeted research into life and habitat of 53 important, scarce and declining species. |
| Notable A | Found in fewer than thirty 10km squares nationally. |
| Notable B | Found in thirty to one hundred 10km squares nationally. |
| DR | Found in 3 or fewer sites in Dorset even if C or L elsewhere. |
| L | Widespread but local |
| C | Ubiquitous, could turn up almost anywhere |

Notes in the final column mostly name food plants but can show more information.

*****

| | | | | | | | | |
|---|---|---|---|---|---|---|---|---|
| 5 | *Micropterix calthella* | | | 37 | Bloxworth | 02-May-05 | | | Tiny, day-flyer May.Feeds on pollen |
| 14 | *Hepialus humuli* | Ghost Moth | | 1 | T. Apple Loft | 13-Aug-94 | C | | Grass roots. Male is white like ghost. |
| 15 | *Hepialus sylvina* | Orange Swift | | 1 | T. Apple Loft | 25-Aug-95 | C | | Roots of bracken, dock etc. |
| 16 | *Hepialus hecta* | Gold Swift | | | T. Barrow Gallop | 07-Jul-00 | L | | Bracken roots |
| 17 | *Hepialus lupulinus* | Common Swift | | 1 | T. Charlie Wight | 31-May-96 | C | | Grass roots. Often abundant |
| 129 | *Incurvaria pectinea* | longhorn | | | T. Long Close | 18-Jun-98 | | | Birch, hazel, hornbeam |
| 130 | *Incurvaria masculella* | | | 1 | T. Ford Pylon | 16-May-04 | | | Sundry leaves, esp. hawthorn |
| 140 | *Nematopogon swammerdamella* | | | 37 | Puddletown | 21-May-03 | | | Large 'longhorn'. Dead leaves |
| 143 | *Nematopogon metaxella* | longhorn | | | T. Apple Loft | 30-Jun-97 | | | Leaf litter |
| 148 | *Nemophora degeerella* | longhorn | | 1 | T. Long Close | 18-Jun-98 | | | Day flyer. Larvae in leaf litter |
| 150 | *Adela reamurella* | longhorn | | 1 | T. Apple Loft | 20-May-95 | | | Day flyer. Larvae in leaf litter |
| 160 | *Phragmataecia castaneae* | Reed Leopard | | 37 | Wareham | 16-Jul-02 | RDB2 | | Dorset and East Anglia only. Common Reed *Phragmites* |
| 161 | *Zeuzera pyrina* | Leopard | | 1 | T. Apple Loft | 21-Jul-96 | C | | Inside various tree branches |
| 162 | *Cossus cossus* | Goat | | 37 | Wareham | 25-Jun-05 | Notable B | | Smells of goat. Larvae up to 5 years inside live oak etc. trees |
| 169 | *Zygaena filipendulae* | Six-spot Burnet | | | T. Charlie Wight | 18-Jul-96 | C | | Birds-foot trefoil |
| 173 | *Apoda limacodes* | Festoon | | 1 | T. Nowhere | 13-Aug-96 | Notable B | | Oak, beech |
| 175 | *Narycia monilifera* | | | | T. Long Close | 18-Jun-98 | | | Lichens |
| 180 | *Diplodoma herminata* | | | | T. Long Close | 18-Jun-98 | | | Lichens |
| 181 | *Taleporia tubulosa* | | | | T. Long Close | 18-Jun-98 | | | Bagworm. Lichens |
| 185 | *Luffia ferchautella* | | | | T. Long Close | 18-Jun-98 | | | Cased larvae on lichens |
| 186 | *Psyche casta* | | | | T. Long Close | 18-Jun-98 | | | Bagworm on grass, lichen |
| 196 | *Morophaga choragella* | | | | T. Apple Loft | 25-Jul-01 | | L | Bracket fungus |
| 203 | *Infurcitinea argentimaculella* | | | | T. Long Close | 18-Jun-98 | | | In silk tubes on lichens |
| 216 | *Nemapogon cloacella* | Cork Moth, micro | | | T. Apple Loft | 05-Jun-02 | | | Bracket fungus |
| 219 | *Nemapogon ruricolella* | | | | T. Long Close | 18-Jun-98 | Notable B | | Bracket fungus |
| 220 | *Nemapogon clematella* | | | | T. Long Close | 07-Aug-98 | | | Hatched ex hazel |
| 224 | *Triaxomera parasitella* | | | | T. Apple Loft | 29-Jun-01 | | | Bracket fungus *coriolus* |
| 227 | *Monopis leavigella* | Skin Moth, micro | | | T. Apple Loft | 16-May-97 | | | In birds nests |
| 228 | *Monopis weaverella* | | | | T. Long Close | 09-May-00 | | | Scavenger on dead animals & faeces |
| 230 | *Monopis crocicapitella* | | | | T. Apple Loft | 15-May-98 | | | Scavenger on dead plants |
| 240 | *Tinea pellionella* | | | | T. Carpenters Workshop | 29-Jun-98 | | | Case-bearing clothes moth on fur, feathers |
| 247 | *Tinea trinotella* | | | | T. Bottom Fish Farm | 17-Aug-96 | | | In birds nests |
| 282 | *Caloptilia elongella* | | | 1 | T. Scout Camp | 18-Jun-98 | | | Leaf mines on Alder |
| 286 | *Caloptilia alchimiella* | | | 1 | T. Scout Camp | 10-Jul-97 | | | Leaf mines on oak |
| 287 | *Caloptilia robustella* | | | 1 | T. Apple Loft | 27-Jul-02 | | | Leaf mines on oak, beech, chestnut |
| 288 | *Caloptilia stigmatella* | | | 2 | T. Apple Loft | 12-Aug-98 | | | Leaf mines on willow |
| 293 | *Caloptilia syringella* | | | | T. Barrow Gallop | 12-Jul-98 | | | Leaf mines on privet, lilac |
| 296 | *Calybites phasianipennella* | | | 1 | T. Apple Loft | 13-Jul-03 | | | Leaf mines on dock, loosestrife, knotgrass |
| 332.1 | *Phyllonorycter leucographella* | Firethorn Leaf Miner | | 1 | T. Apple Loft | 25-Jan-02 | Naturalised | | Immigrant 1989 on *Pyracantha* , sighting is of overwintering larva |
| 343 | *Phyllonorycter quinnata* | | | | T. Scout Camp | 18-Jun-98 | | L | Leaf miner on hornbeam |
| 373 | *Synanthedon tipuliformis* | Currant Clearwing | | 1 | T. Apple Loft | 05-Jun-01 | Notable B | | Blackcurrant bushes |
| 377 | *Synanthedon flaviventris* | Sallow Clearwing | | | T. Duck Pond Heath | 28-Jun-02 | Notable B | | 2 yr cycle, only hatches even years |
| 385 | *Anthophila fabriciana* | Nettle-tap, micro | | 1 | T. Apple Loft | 02-Jun-97 | | | Nettle |
| 410 | *Argyresthia brockeella* | | | | T. Roundabout | 09-Aug-96 | | | Catkins of birch, alder |
| 416 | *Argyresthia glaucinella* | | | | T. Scout Camp | 18-Jun-98 | | | Oak bark |

| 424 | *Yponomeuta evonymella* | Bird-cherry Ermine, micro | 1 | T. Apple Loft | 28-Jul-95 | | Bird cherry |
|-----|------|------|---|------|------|------|------|
| 425 | *Yponomeuta padella* | Orchard Ermine, micro | | T. Charlie Wight | 22-Jul-96 | | Blackthorn, hawthorn in web |
| 428 | *Yponomeuta rorrella* | Willow Ermine, micro | | T. Charlie Wight | 22-Jul-96 | pRDB3 | Webs on white willow |
| 431 | *Yponomeuta sedella* | | | T. Barrow Gallop | 28-Jul-00 | Notable B | Webs on orpine |
| 435 | *Zelleria hepariella* | | | T. Badgers Alley | 29-Mar-98 | | On ash in spun leaves |
| 436 | *Pseudoswammerdamia combinella* | | 37 | Studland | 19-May-04 | | On blackthorn |
| 449 | *Prays fraxinella* | Ash Bud Moth, micro | | T. Duck Ponds | 29-Jun-97 | | Ash buds |
| 451 | *Ypsolopha mucronella* | | | T. Apple Loft | 16-Mar-96 | | Spindle, hibernates |
| 452 | *Ypsolopha nemorella* | | 1 | T. Buddens | 14-Jun-97 | | Honeysuckle |
| 453 | *Ypsolopha dentella* | Honeysuckle Moth, micro | | T. Ford | 13-Jul-97 | | Honeysuckle |
| 455 | *Ypsolopha scabrella* | | 37 | Durlston | 02-Aug-04 | | Gardens, woods on apple or hawthorn |
| 458 | *Ypsolopha alpella* | | 2 | T. Apple Loft | 07-Aug-03 | | Web on oak leaves |
| 460 | *Ypsolopha parenthesella* | | 2 | T. North Pond | 07-Aug-01 | | Oak, hazel |
| 461 | *Ypsolopha ustella* | | 2 | T. Ford | 12-Oct-96 | | Oak, overwinters as adult |
| 462 | *Ypsolopha sequella* | | 2 | T. Apple Loft | 02-Aug-01 | | Field maple, sycamore |
| 464 | *Plutella xylostella* | Diamond-back Moth, micro | | T. Ford | 07-Apr-96 | Migrant | Common, travels vast distances, often abundant |
| 465 | *Plutella porrectella* | | | T. Hyde Heath | 18-Jun-97 | | Dames violet |
| 470 | *Orthotaelia sparganella* | | | T. Apple Loft | 28-Jul-01 | | Ponds etc on bur-reed |
| 472 | *Digitivalva pulicariae* | | | T. Ford | 01-May-97 | | Fleabane feeder |
| 493 | *Coleophora serratella* | | | T. Scout Camp | 18-Jun-98 | | Birch, alder |
| 512 | *Coleophora binderella* | | | T. Scout Camp | 18-Jun-98 | | Mines on alder, hornbeam, birch, hazel |
| 544 | *Coleophora albicosta* | | 1 | T. Lot 1 | 30-May-97 | | Feeds on and in seeds of gorse |
| 607 | *Elachista canapennella* | | | T. Apple Loft | 24-Aug-03 | | Mines in grass blades |
| 640 | *Batia lunaris* | | | T. Apple Loft | 15-Jul-03 | | Feeds under tree bark |
| 641 | *Batia lambdella* | | | T. Scout Camp | 10-Jul-97 | | Similar to above but larger. Gorse stems |
| 642 | *Batia unitella* | | | T. Ford | 13-Jul-96 | | Fungus under dead bark |
| 644 | *Borkhausenia fuscescens* | | | T. Apple Loft | 10-Jul-98 | | Dead leaves, birds nests |
| 646 | *Telechrysis tripuncta* | | | T. South Long Close | 18-Jun-98 | | Dead wood |
| 647 | *Hofmannophila pseudospretella* | Brown House Moth, micro | | T. Apple Loft | 29-Jun-96 | | Sundry detritus |
| 648 | *Endrosis sarcitrella* | White-shouldered House Moth | | T. Apple Loft | 20-Apr-96 | | Continuously brooded on dried plant matter etc |
| 649 | *Esperia sulphurella* | | | T. Badgers Alley | 12-May-98 | | Dead wood |
| 652 | *Alabonia geoffrella* | | 37 | Langton | 27-May-01 | | Spectacular day-flyer. Feeds inside bramble stems |
| 654 | *Pleurota bicostella* | | 2 | T. Charlie Wight | 14-Jul-97 | | Shoots of erica (bell heather) |
| 656 | *Tachystola acroxantha* | | 2 | T. Apple Loft | 19-Sep-02 | pRDB3 | Leaf litter |
| 658 | *Carcina quercana* | | | T. Charlie Wight | 05-Aug-96 | | Oak, beech |
| 663 | *Diurnea fagella* | | 2 | T. South Lodge | 12-Apr-96 | | Flies in Feb. Deciduous trees |
| 688 | *Agonopterix heracliana* | | | T. Charlie Wight | 09-Oct-96 | | Umbellifers |
| 689 | *Agonopterix ciliella* | | | T. Apple Loft | 09-Apr-96 | | Umbellifers |
| 692 | *Agonopterix subpropinquella* | | | T. Apple Loft | 02-Apr-97 | | Knapweed, thistle. Hibernates over winter after emerging August on |
| 695 | *Agonopterix alstromeriana* | | 2 | T. Apple Loft | 24.04.03 | | Hemlock |
| 697 | *Agonopterix arenella* | | | T. Apple Loft | 15-Apr-95 | | Thistles, knapweed. Hibernates |
| 700 | *Agonopterix pallorella* | | 37 | Durlston | 02-Aug-04 | Notable B | Knapweed, saw wort in rolled leaves. Overwinters |
| 701 | *Agonopterix ocellana* | | | T. Fish Farm | 24-Mar-96 | | Spun leaves of willow. Hibernates |
| 705 | *Agonopterix umbellana* | | | T. Barrow Gallop | 01-Apr-99 | | Gorse in silk tubes. Hibernates |
| 706 | *Agonopterix nervosa* | | 37 | Studland | 19-Aug-00 | | Spun shoots of broom, gorse |

| | | | | | | | |
|---|---|---|---|---|---|---|---|
| 714 | *Agonopterix yeatiana* | | 2 | T. Ford | 07-Apr-96 | | Umbellifers inc. wild carrot & hemlock water-dropwort. Hibernates |
| 726 | *Metzneria metzneriella* | | 37 | Portland | 24-Jun-05 | | Dry limestone pastures. Knap weed, saw wort |
| 727.1 | *Metzneria aprilella* | | 37 | Cashmoor | 17-Jun-05 | Notable A | Chalk downs |
| 728 | *Monochroa cytisella* | | | T. Dynamite | 31-Jul-02 | | Bracken |
| 729 | *Isophrictis striatella* | | 37 | Stanpit | 29-Jul-05 | | Feeds on tansy |
| 752 | *Aristotelia ericinella* | | | T. Apple Loft | 29-Jul-97 | | Day flying,  on heather |
| 762 | *Athrips mouffetella* | | | T.Barrow Gallop | 21-Jul-00 | | Woods and gardens, on honeysuckle |
| 770 | *Carpatolechia proximella* | | | T. Brick Kiln | 25-May-98 | | Birch, alder |
| 771 | *Teleiodes alburnella* | | | T. Ford | 18-Jun-98 | | Birch |
| 774 | *Teleiodes luculella* | | | T. Brick Kiln | 31-May-97 | | Oak woods |
| 776 | *Teleiopsis diffinis* | | | T. Brick Kiln | 31-May-97 | | Sheep's sorrel |
| 779 | *Bryotropha affinis* | | | T. Apple Loft | 17-Jun-02 | | Moss on walls |
| 789 | *Bryotropha domestica* | | | T. Barrow Gallop | 01-Oct-98 | | Moss on walls, common in south |
| 796 | *Aroga velocella* | | | T. Apple Loft | 22-Jul-02 | | Sheep's sorrel |
| 797 | *Neofaculta ericetella* | | | T. Charlie Wight | 27-May-00 | | Common on heather |
| 819 | *Scrobipalpa costella* | | | T. Scout Camp | 20-Feb-98 | | Damp woods on bittersweet |
| 841 | *Sophronia semicostella* | | | T. Barrow Gallop | 08-Jul-00 | | ?Sweet vernal grass |
| 844 | *Syncopacma larseniella* | | | T. Apple Loft | 01-Jun-00 | Notable B | In spun leaves of bird's foot trefoil |
| 851 | *Dichomeris alacella* | | | T. Apple Loft | 19-Aug-96 | Notable B | 1st Dorset record. Possibly on tree lichens |
| 854 | *Anacampsis blattariella* | | 2 | T. Ford | 13-Jul-96 | | Birch |
| 856 | *Anarsia spartiella* | | | T. Apple Loft | 14-Jul-00 | | Gorse, broom, Dyer's greenweed |
| 866 | *Brachmia blandella* | | | T. Ford | 13-Jul-96 | | Large old gorse |
| 868 | *Helcystogramma rufescens* | | | T. Ford | 13-Jul-96 | | Grasses |
| 870 | *Oegoconia quadripuncta* | | | T. Apple Loft | 09-Jul-98 | | South UK on leaf litter |
| 873 | *Blastobasis lignea* | | | T. Bottom Fish Farm | 17-Aug-96 | | Introduced. Decaying vegetable matter |
| 874 | *Blastobasis decolorella* | | 2 | T. Apple Loft | 12-Jun-01 | | Introduced. Decaying vegetable matter |
| 886 | *Mompha ochraceella* | | | T. Ford | 13-Jul-96 | | Mines on great willowherb |
| 898 | *Limnaecia phragmitella* | | | T. Roundabout | 09-Aug-96 | | Inside bulrush seed heads |
| 921 | *Phtheochroa inopiana* | | | T. Apple Loft | 08-Jul-99 | | Damp woods, common fleabane |
| 924 | *Hysterophora maculosana* | | | T. Bottom Fish Farm | 05-May-97 | | Woodland, in bluebell seeds |
| 930 | *Gynnidomorpha alismana* | | | T. Roundabout | 09-Aug-96 | Notable B | Water plantain |
| 936 | *Cochylimorpha straminea* | | 2 | T. Apple Loft | 26-Aug-02 | | Common knapweed |
| 937 | *Agapeta hamana* | | | T. Apple Loft | 20-Jun-94 | | Thistle |
| 938 | *Agapeta zoegana* | | 2 | T. Nowhere | 13-Aug-96 | | Common knapweed |
| 945 | *Aethes cnicana* | | | T. Duck Pond Heath | 05-Jul-97 | | Various thistle seeds |
| 946 | *Aethes rubigana* | | | T. Apple Loft | 16-Jul-95 | | Burdock seeds |
| 950 | *Aethes francillana* | | 37 | Durlston | 02-Aug-04 | | Wild carrot, coastal limestone downs |
| 951 | *Aethes beatricella* | | | T. Ford | 13-Jul-96 | | South-east England on hemlock |
| 954 | *Eupoecilia angustana* | | | T. Apple Loft | 29-Jul-97 | | Plantain, yarrow, heather |
| 965 | *Cochylis hybridella* | | 37 | Portland | 24-Jun-05 | | Chalk & limestone in south.Ox tongue & hawk's beard |
| 966 | *Cochylis atricapitana* | | 2 | T. Bottom Fish Farm | 17-Aug-96 | | Usually coastal, chalk land. Ragwort |
| 968 | *Cochylis nana* | | | T. Brick Kiln | 31-May-97 | | Birch catkins |
| 969 | *Pandemis corylana* | Chequered Fruit-tree Tortrix, mic | | T. Ford | 20-Aug-96 | | Deciduous trees |
| 970 | *Pandemis cerasana* | Barred Fruit-tree Tortrix, micro | | T. Ford | 13-Jul-96 | | Woods, trees especially fruit |
| 972 | *Pandemis heparana* | Dark Fruit-tree Tortrix, micro | | T. Apple Loft | 20-Jun-94 | | Woods, trees especially fruit, in rolled leaves |

| | | | | | | | |
|---|---|---|---|---|---|---|---|
| 974 | *Argyrotaenia ljungiana* | | | T. Charlie Wight | 05-Aug-96 | | On marsh gentian, bog myrtle, heather. |
| 976 | *Archips oporana* | | 2 | T. Barrow Gallop | 21-Jul-00 | pRDB1 | Scots pine |
| 977 | *Archips podana* | Large Fruit-tree Tortrix, micro | 2 | T. Apple Loft | 06-Jul-94 | | Various deciduous trees. Sexually dimorphic |
| 980 | *Archips xylosteana* | | 38 | Verwood | 15-Jun-03 | | Trees and shrubs |
| 985 | *Cacoecimorpha pronubana* | Carnation Tortrix, micro | | T. Apple Loft | 31-Jul-97 | | Sundry plants |
| 986 | *Syndemis musculana* | | | T. Brick Kiln | 31-May-97 | | Woods & moors. Deciduous trees |
| 987 | *Ptycholomoides aeriferanus* | | | T. Charlie Wight | 22-Jul-96 | | Arrived UK 1951. Spun larch needles |
| 993 | *Clepsis spectrana* | | 2 | T. Margarets Garden | 07-Jul-96 | | Damp woods |
| 998 | *Epiphyas postvittana* | Light Brown Apple Moth, micro | 2 | T. Apple Loft | 08-Oct-96 | Naturalised | Australian immigrant in 1930. Eats anything, seen all year round |
| 1001 | *Lozotaeniodes formosanus* | | | T. Ford | 13-Jul-96 | | Southern UK on scots pine |
| 1002 | *Lozotaenia forsterana* | | | T. Apple Loft | 01-Jun-95 | | Large tortrix, suburban, shrubs esp. ivy |
| 1006 | *Epagoge grotiana* | | 2 | T. Scout Camp | 18-Jun-98 | | Widespread, oak, hawthorn |
| 1007 | *Capua vulgana* | | | T. Ford | 12-Jun-02 | | Woods, alder, rowan, bilberry |
| 1010 | *Ditula angustiorana* | Red-barred Tortrix, micro | | T. Ford | 13-Jul-96 | | Various trees, shrubs |
| 1011 | *Pseudargyrotoza conwagana* | | | T. Lot 1 | 12-Jun-97 | | Small colourful tortrix, mainly ash, privet |
| 1012 | *Sparganothis pilleriana* | | | T. Nowhere | 12-Sep-96 | Notable B | Polyphagous (eats many different kinds of food) |
| 1020 | *Cnephasia stephensiana* | Grey Tortrix, micro | | T. Ford | 13-Jul-96 | | Deciduous trees |
| 1025 | *Tortricodes alternella* | | 2 | T. Bottom Fish Farm | 24-Mar-96 | | Flies in February. Rolled leaves oak, hornbeam |
| 1030 | *Eana incanana* | | | T. Apple Loft | 02-Jul-97 | | Flowers of bluebell, ox-eye daisy |
| 1032 | *Aleimma loeflingiana* | | | T. Margarets Garden | 07-Jul-96 | | Woodland on oak |
| 1033 | *Tortrix viridana* | Green Oak Tortrix, micro | 2 | T. Hyde Heath | 18-Jun-97 | | Abundant on oak |
| 1036 | *Acleris forsskaleana* | | 38 | Durlston | 02-Aug-04 | | Field maple, sycamore |
| 1038 | *Acleris laterana* | | | T. Charlie Wight | 22-Jul-96 | | Roses |
| 1041 | *Acleris sparsana* | | | T. Apple Loft | 12-Nov-01 | | Beech, sycamore |
| 1045 | *Acleris notana* | | 2 | T. Ford | 08-Nov-98 | | Birch |
| 1047 | *Acleris schalleriana* | | | T. Apple Loft | 30-Nov-97 | | Wayfaring tree, guelder rose |
| 1048 | *Acleris variegana* | Garden Rose Tortrix, micro | 2 | T. Apple Loft | 15-Aug-96 | | Roses etc |
| 1053 | *Acleris hastiana* | | | T. Ford | 13-Jul-96 | | Many forms. Small-leaved willows |
| 1054 | *Acleris cristana* | (See W B L Manley guide 1973) | | T. Badgers Alley | 12-May-98 | | Variable, 119 named forms! Haw- blackthorn |
| 1061 | *Acleris literana* | | 2 | T. Apple Loft | 07-Mar-94 | | Oak woods, hibernates |
| 1062 | *Acleris emargana* | | 38 | Studland | 14-Sep-02 | | Leaves, shoots of sallow, poplar, birch |
| 1063 | *Celypha striana* | | 2 | T. Apple Loft | 16-Jul-95 | | Grassland on dandelion roots |
| 1076 | *Celypha lacunana* | | | T. Buddens | 11-Jun-96 | | Very common, herbaceous plants |
| 1079 | *Piniphila bifasciana* | | | T. Ford | 13-Jul-96 | | Shoots and male flowers of scots pine |
| 1082 | *Hedya pruniana* | Plum Totrix, micro | 2 | T. Ford | 13-Jul-96 | | Prunus, inc hawthorn and plum |
| 1086 | *Hedya salicella* | | | T. Scout Camp | 10-Jul-97 | | Sallow |
| 1092 | *Apotomis turbidana* | | 2 | T. Ford | 13-Jul-96 | | Spun birch leaves |
| 1093 | *Apotomis betuletana* | | | T. Brick Kiln | 16-Jul-96 | | Spun birch leaves |
| 1104 | *Endothenia quadrimaculana* | | | T. Margaret's Garden | 07-Jul-96 | | Upper parts of roots of marsh woundwort |
| 1111 | *Bactra lancealana* | | 2 | T. Bottom Fish Farm | 17-Aug-96 | | Variable size. Marsh & heaths on rushes |
| 1115 | *Ancylis achatana* | | | T. Apple Loft | 12-Jul-97 | | Spun leaves of hawthorn, blackthorn |
| 1120 | *Ancylis mitterbacheriana* | | 2 | T. Lot 1 | 30-May-97 | | Oak, beech |
| 1132 | *Epinotia subocellana* | | | T. Lot 1 | 30-May-97 | | Sallows in damp woods |
| 1133 | *Epinotia bilunana* | | 3 | T. Ford | 13-Jul-96 | | Birch catkins internally through winter |
| 1134 | *Epinotia ramella* | | 3 | T. Ford | 12-Oct-96 | | Birch, sallow catkins |

| 1135 | *Epinotia demarniana* | | | T. Ford | 08-Jun-97 | Notable B | Birch, alder, sallow catkins |
|------|----------------------|---|---|---------|-----------|-----------|------------------------------|
| 1137 | *Epinotia tetraquetrana* | | | T. Bottom Fish Farm | 05-May-97 | | Birch, alder in folded leaf |
| 1138 | *Epinotia nisella* | | 3 | T. Charlie Wight | 05-Aug-96 | | Variable. On sallow, poplar |
| 1155 | *Epinotia brunnichana* | | | T. Charlie Wight | 05-Aug-96 | | Rolled leaf of birch, hazel or sallow |
| 1156 | *Epinotia solandriana* | | | T. Dynamite | 31-Jul-02 | | Rolled leaf of birch, hazel or sallow. Variable |
| 1157 | *Crocidosema plebejana* | | | T. Ford | 12-Oct-96 | Notable A | Tree mallow |
| 1165 | *Zeiraphera isertana* | | | T. Bottom Fish Farm | 01-Aug-97 | | Folded or spun oak leaves |
| 1169 | *Gypsonoma dealbana* | | | T. Ford | 13-Jul-96 | | Deciduous trees |
| 1174 | *Epiblema cynosbatella* | | | T. Apple Loft | 08-Jun-96 | | Roses |
| 1175 | *Epiblema uddmanniana* | Bramble Shoot Moth, micro | | T. Apple Loft | 28-Jun-94 | | Brambles |
| 1177 | *Epiblema rosaecolana* | | | T. Ford | 13-Jul-96 | | Wild roses |
| 1183 | *Epiblema foenella* | | | T. Ford | 13-Jul-96 | | Mugwort |
| 1184 | *Epiblema scutulana* | | 3 | T. Dynamite | 20-Jul-02 | | Spear or musk thistle in stems or roots |
| 1197 | *Eucosma campoliliana* | | | T. Ford | 19-Jun-96 | | Leaves or stems of ragwort |
| 1201 | *Eucosma cana* | | 3 | T. Ford | 13-Jul-96 | | Thistle or black knapweed |
| 1204 | *Thiodia citrana* | | 3 | T. Apple Loft | 12-Jul-97 | | Various composite flowers, inc. yarrow |
| 1209 | *Pseudococcyx turionella* | Pine Bud Moth, micro | | T. Charlie Wight | 31-May-96 | | Scots pine shoots, buds |
| 1211 | *Rhyacionia pinicolana* | | 3 | T. Margaret's Garden | 18-Aug-96 | | Scots pine shoots, buds |
| 1212 | *Rhyacionia pinivorana* | Spotted Shoot, micro | 3 | T. Lot 1 | 30-May-97 | | Scots pine shoots, buds |
| 1219 | *Lathronympha strigana* | | | T. Brick Kiln | 31-May-97 | | Open woods on St.John's wort |
| 1229 | *Pammene albuginana* | | | T. Carpenter's Workshop | 05-Jul-96 | Notable B | Spongy gall oak |
| 1236 | *Pammene fasciana* | | | T. Ford | 13-Jul-96 | Notable B | Inside acorns of oak, nuts of sweet chestnut |
| 1255 | *Cydia succedana* | | | T. Charlie Wight | 19-Oct-03 | | Seeds of gorse, broom |
| 1259 | *Cydia fagiglandana* | | | T. Bottom Fish Farm | 17-Aug-96 | | Feeds inside beech nuts |
| 1260 | *Cydia splendana* | | | T. Roundabout | 09-Aug-96 | | Feeds inside acorns or sweet chestnuts |
| 1261 | *Cydia pomonella* | Codling Moth, micro | | T. Charlie Wight | 05-Aug-96 | | Inside fruits of apple, pear etc. |
| 1262 | *Cydia amplana* | | | T. Apple Loft | 19-Aug-96 | Vagrant | Scarce migrant, cork feeder |
| 1272 | *Pammene aurana* | | 38 | Cashmoor | 17-Jun-05 | | Feed on hogweed seeds spun together |
| 1273 | *Dichrorampha petiverella* | | 3 | T. Apple Loft | 18-Jun-05 | | Day flyer on rough land. Larvae feed in roots of yarrow & tansy |
| 1288 | *Alucita hexadactyla* | Twenty-plume Moth | 3 | T. Apple Loft | 06-Oct-93 | | Honeysuckle. Flies all year round |
| 1290 | *Chilo phragmitella* | pyrale | | T. Apple Loft | 27-Jun-96 | | Common reed stems |
| 1292 | *Calamotropha paludella* | pyrale | 3 | T. Ford | 13-Jul-96 | Notable B | Dead bulrush stems |
| 1293 | *Chrysoteuchia culmella* | Garden Grass-veneer, pyrale | | T. Ford | 13-Jul-96 | | Abundant on grassland |
| 1294 | *Crambus pascuella* | pyrale | | T. Brick Kiln | 02-Jun-96 | | Very common, food unknown! |
| 1296 | *Crambus silvella* | pyrale | | T. Charlie Wight | 05-Aug-96 | pRDB3 | Sedges on boggy heathland |
| 1299 | *Crambus hamella* | pyrale | 3 | T. Ford | 23-Aug-97 | Notable B | Food unknown. Dry heaths |
| 1302 | *Crambus perlella* | pyrale | 3 | T. Ford | 13-Jul-96 | | Grasses |
| 1303 | *Agriphila selasella* | pyrale | 3 | T. Apple Loft | 28-Jul-95 | | Saltmarshes |
| 1304 | *Agriphila straminella* | pyrale | | T. Ford | 13-Jul-96 | | Abundant on grassland |
| 1305 | *Agriphila tristella* | pyrale | 3 | T. Roundabout | 09-Aug-96 | | Abundant on grassland |
| 1306 | *Agriphila inquinatella* | pyrale | | T. Charlie Wight | 05-Aug-96 | | Small grasses sheep's fescue |
| 1307 | *Agriphila latistria* | pyrale | 3 | T. Apple Loft | 15-Aug-96 | | Dry heaths |
| 1309 | *Agriphila geniculea* | pyrale | 3 | T. Bottom Fish Farm | 17-Aug-96 | | Dry pasture |
| 1313 | *Catoptria pinella* | pyrale | | T. Apple Loft | 11-Jul-95 | | Heathy, boggy woodlands |
| 1316 | *Catoptria falsella* | pyrale | 3 | T. Apple Loft | 16-Jul-95 | | Mosses |

| 1329 | *Donacaula forficella* | pyrale | 3 | T. Margaret's Garden | 07-Jul-96 | | Common reed. Larva cuts a leaf to use as raft to float between plants |
|------|------|------|------|------|------|------|------|
| 1330 | *Donacaula mucronellus* | pyrale | | T. Ford | 13-Jul-96 | | Common reed, *Phragmites* |
| 1331 | *Acentria ephemerella* | Water Veneer, pyrale | | T. Nowhere | 13-Aug-96 | | Aquatic larvae down to 2 metres |
| 1333 | *Scoparia pyralella* | pyrale | 3 | T. Ford Meadow | 19-Jun-96 | | Decaying plant material |
| 1334 | *Scoparia ambigualis* | pyrale | | T. South Lodge | 01-Jun-96 | | Common but food unknown! |
| 1334.1 | *Scoparia basistrigalis* | pyrale | | T. Bottom Fish Farm | 17-Aug-96 | | Common but food unknown! |
| 1338 | *Dipleurina lacustrata* | pyrale | | T. Ford | 13-Jul-96 | | Moss |
| 1340 | *Eudonia truncicolella* | pyrale | | T. Apple Loft | 17-Jul-98 | | Moss |
| 1342 | *Eudonia angustea* | pyrale | | T. Apple Loft | 11-Oct-96 | | Moss |
| 1343 | *Eudonia delunella* | pyrale | | T. Ford | 13-Jul-96 | Notable B | Lichen/moss |
| 1344 | *Eudonia mercurella* | pyrale | | T. Bottom Fish Farm | 17-Aug-96 | | Moss |
| 1345 | *Elophila nymphaeata* | Brown China-mark, pyrale | 4 | T. Apple Loft | 01-Jul-95 | | Aquatic larvae on pondweeds, bur-reed etc. |
| 1348 | *Parapoynx stratiotata* | Ringed China-mark, pyrale | 4 | T. Apple Loft | 22-Jul-02 | | Aquatic larvae on pondweeds, elodea etc. |
| 1350 | *Nymphula nitidulata* | Beautiful China Mark, pyrale | 4 | T. Apple Loft | 01-Jul-95 | | Aquatic larvae on lilies, bur-reed etc. |
| 1354 | *Cataclysta lemnata* | Small China-mark, pyrale | | T. Apple Loft | 17-Aug-97 | | Sexually dimorphic. Duckweed |
| 1356 | *Evergestis forficalis* | Garden Pebble, pyrale | | T. Apple Loft | 04-Jul-95 | | Garden pest on brassicas |
| 1356.1 | *Evergestis limbata* | pyrale | 4 | T. Apple Loft | 18-Jul-02 | Vagrant | Scarce migrant, garlic mustard |
| 1358 | *Evergestis pallidata* | pyrale | | T. Apple Loft | 12-Aug-96 | | Damp, open woodland. Crucifers |
| 1361 | *Pyrausta aurata* | pyrale | | T. Apple Loft | 15-Aug-96 | | On mint |
| 1362 | *Pyrausta purpuralis* | pyrale | 4 | T. Apple Loft | 09-Jul-97 | | Mint and thyme |
| 1363 | *Pyrausta ostrinalis* | pyrale | | T. Apple Loft | 11-Aug-97 | | Food unknown |
| 1365 | *Pyrausta despicata* | pyrale | | T. Apple Loft | 05-Sep-03 | | Plantain |
| 1367 | *Pyrausta cingulata* | pyrale | 38 | Durlston | 29-Jul-04 | | Coastal, limestone, on wild thyme. Double-brooded |
| 1368 | *Loxostege sticticalis* | pyrale | | T. Apple Loft | 09-Aug-97 | Migrant | Scarce migrant, mugwort |
| 1370 | *Sitochroa palealis* | pyrale | | T. Apple Loft | 17-Aug-96 | Notable B | Coastal, local, fennel |
| 1375 | *Ostrinia nubilalis* | European Corn-borer, pyrale | 4 | T. Apple Loft | 12-Sep-99 | Migrant | Maize, mugwort |
| 1376 | *Eurrhypara hortulata* | Small Magpie, pyrale | 4 | T.Ford | 13-Jul-96 | | Nettle |
| 1377 | *Perinephela lancealis* | pyrale | 4 | T. Brick Kiln | 24-May-97 | | Hemp agrimony, damp woods and marshes |
| 1378 | *Phlyctaenia coronata* | pyrale | 4 | T. Apple Loft | 11-Jul-95 | | Larvae on elder leaves, hibernate over winter, pupate in spring |
| 1380 | *Phlyctaenia perlucidalis* | pyrale | 4 | T. Apple Loft | 25-Jun-01 | DR | Fens, perhaps from Poole harbour |
| 1382 | *Anania verbascalis* | pyrale | 4 | T. Apple Loft | 26-Jun-97 | Notable B | Wood sage. Larvae eat stems causing plant to wilt |
| 1383 | *Psammotis pulveralis* | pyrale | 4 | T. Apple Loft | 12-Jul-05 | Migrant | Scarce, in marshes |
| 1385 | *Ebulea crocealis* | pyrale | 4 | T. Bottom Fish Farm | 01-Aug-97 | | Common fleabane, ploughman's spikenard |
| 1389 | *Udea fulvalis* | pyrale | 4 | T. Apple Loft | 16-Jul-01 | Migrant | Occasional immigrant to South Dorset. Mint |
| 1390 | *Udea prunalis* | pyrale | 4 | T. Apple Loft | 16-Jul-01 | | Blackthorn thickets |
| 1392 | *Udea olivalis* | pyrale | | T. Apple Loft | 01-Jul-95 | | Herbaceous plants |
| 1395 | *Udea ferrugalis* | pyrale | 4 | T. Ford | 13-Jul-96 | Migrant | Common |
| 1397 | *Mecyna asinalis* | pyrale | 38 | Durlston | 29-Jul-04 | Notable B | Southern UK coasts on wild madder |
| 1398 | *Nomophila noctuella* | Rush Veneer, pyrale | 4 | T. Apple Loft | 08-Sep-94 | Migrant | Common |
| 1399 | *Dolicarthria punctalis* | pyrale | 38 | Durlston | 29-Jul-04 | Notable B | Only on SW UK coasts. Decaying plant matter |
| 1405 | *Pleuroptya ruralis* | Mother of Pearl, pyrale | 4 | T. Apple Loft | 11-Jul-95 | | Nettle |
| 1408 | *Palpita vitrealis* | pyrale | 38 | Durlston | 28-Oct-05 | Migrant | Scarce. On jasmine and olive |
| 1413 | *Hypsopygia costalis* | Gold Triangle, pyrale | 4 | T. Apple Loft | 29-Jul-94 | | Hay barns, squirrels dreys |
| 1414 | *Synaphe punctalis* | pyrale | | T. Apple Loft | 16-Jul-95 | Notable B | Hypnum moss. Local coastal |
| 1415 | *Orthopygia glaucinalis* | pyrale | | T. Apple Loft | 14-Jul-95 | | Decaying plant material |

| 1424 | *Endotricha flammealis* | pyrale | 4 | T. Ford | 13-Jul-96 | | Decaying plant material |
|---|---|---|---|---|---|---|---|
| 1425 | *Galleria mellonella* | Wax Moth, pyrale | 5 | T. Apple Loft | 17-Oct-95 | | Honeycomb, larvae in beehives |
| 1428 | *Aphomia sociella* | Bee moth, pyrale | 5 | T. Apple Loft | 06-Jun-03 | | Honeycomb, larvae live in beehives |
| 1435 | *Conobathra tumidana* | pyrale | 38 | Durlston | 14-Aug-04 | Migrant | Scarce, local, probably migrant |
| 1437 | *Acrobasis consociella* | pyrale | | T. Ford | 13-Jul-96 | | Gregarious on oak |
| 1438 | *Trachycera suavella* | pyrale | | T. Charlie Wight | 22-Jul-96 | | Blackthorn thickets |
| 1439 | *Trachycera advenella* | pyrale | | T. Ford | 13-Jul-96 | | Hawthorn |
| 1440 | *Trachycera marmorea* | pyrale | | T. Hyde Heath | 18-Jun-97 | | Blackthorn |
| 1441 | *Oncocera semirubella* | pyrale | 5 | T. Apple Loft | 24-Sep-05 | Notable B | Web in legumes. Coastal limestone |
| 1442 | *Pempelia palumbella* | pyrale | 5 | T. Brick Kiln | 16-Jul-96 | | Heathers |
| 1443 | *Pempelia genistella* | pyrale | | T. Ford | 13-Jul-96 | Notable A | Gorse. Very local on heaths |
| 1451 | *Pyla fusca* | pyrale | | T. Ford | 03-Jun-97 | | Heather |
| 1452 | *Phycita roborella* | pyrale | 5 | T. Brick Kiln | 16-Jul-96 | | Oak |
| 1454 | *Dioryctria abietella* | pyrale | | T. Apple Loft | 16-Jul-97 | Naturalised | Scots pine |
| 1454.2 | *Dioryctria sylvestrella* | pyrale | 5 | T. Apple Loft | 28-Jul-00 | Naturalised | 1st Dorset record.  Immigrant on Scots pine |
| 1455 | *Dioryctria simplicella* | pyrale | | T. Barrow Gallop | 14-Jul-00 | | Scots pine shoots |
| 1457 | *Hypochalcia ahenella* | pyrale | 38 | Cashmoor | 17-Jun-05 | | Chalk downs in southern England |
| 1458 | *Myelois circumvoluta* | Thistle Ermine, pyrale | 5 | T. Apple Loft | 27-Jun-94 | | Eats thistles |
| 1461 | *Assara terebrella* | pyrale | 38 | Arne | 14-Jun-03 | Naturalised | Feeds on seeds inside cones of Norway spruce |
| 1465 | *Nephopterix angustella* | pyrale | 38 | Durlston | 10-Oct-05 | Notabla B | In spindle berries. Moth has distinctive black tufts |
| 1470 | *Euzophera pinguis* | pyrale | 5 | T. Bottom Fish Farm | 17-Aug-96 | | Ash bark |
| 1474 | *Ephestia parasitella* | pyrale | 5 | T. Apple Loft | 28-May-04 | | On dead plant matter |
| 1483 | *Phycitodes binaevella* | pyrale | | T. Apple Loft | 30-Jun-00 | | Spear thistle on light soils |
| 1484 | *Phycitodes saxicola* | pyrale | | T. Apple Loft | 17-Jun-02 | | Coastal |
| 1485 | *Phycitodes maritima* | pyrale | 5 | T. Apple Loft | 18-Aug-98 | | Coastal, yarrow and ragwort |
| 1488 | *Agdistis bennetii* | plume | 5 | T. Apple Loft | 19-Aug-96 | | Sea lavender on salt marshes |
| 1493 | *Buckleria paludum* | plume | | T. Nowhere | 13-Aug-96 | pRDB3 | Round-leaved sundew |
| 1494 | *Capperia britanniodactyla* | plume | 5 | T. Dynamite | 25-Jul-03 | Notable B | Wood sage stems |
| 1495 | *Marasmarcha lunaedactyla* | plume | 5 | T. Apple Loft | 03-Jul-06 | | Dry downs and coasts, on rest harrow |
| 1496 | *Cnaemidophorus rhododactyla* | plume | | T. Apple Loft | 04-Jul-01 | pRDB2 | 1st Dorset record. Probable migrant. Breeds in Kent & Essex. Dog rose |
| 1497 | *Amblyptilia acanthadactyla* | plume | 5 | T. Apple Loft | 16-Oct-96 | | Indoors, hibernates |
| 1501 | *Platyptilia gonodactyla* | plume | | T. Apple Loft | 30-May-99 | | Coltsfoot |
| 1503 | *Platyptilia ochrodactyla* | plume | 38 | Stanpit | 29-Jul-05 | Notable B | Feeds in tansy stems |
| 1504 | *Platyptilia pallidactyla* | plume | 38 | Holnest | 09-Jul-05 | | Dry sandy ground, on yarrow |
| 1507 | *Stenoptilia zophodactylus* | plume | 5 | T. Wembere Heath | 30-Sep-98 | | Marsh gentian, centuary and yellow wort |
| 1513 | *Pterophorus pentadactyla* | White Plume | 5 | T. Apple Loft | 12-Jul-94 | | On bindweed *Convolvulus*, often seen in gardens |
| 1517 | *Adaina microdactyla* | plume | 5 | T. Bottom Fish Farm | 01-Aug-97 | | Hemp agrimony |
| 1519 | *Euleioptilus carphodactyla* | plume | 38 | Seacombe | 09-Aug-03 | Notable B | Scarce on dry limestone soils. Ploughman's spikenard |
| 1523 | *Oidaematophorus lithodactyla* | plume | 38 | King's Stag | 13-Jul-03 | | On common fleabane & ploughman's spikenard |
| 1524 | *Emmelina monodactyla* | plume | 5 | T. Apple Loft | 09-Sep-95 | | On bindweed *Convolvulus*, spring & autumn, often in gardens |
| 1631 | *Poecilocampa populi* | December Moth | 6 | T. Apple Loft | 03-Dec-93 | C | Deciduous trees |
| 1633 | *Eriogaster lanestris* | Small Eggar | 37 | Shapwick | 13-Apr-06 | Notable B | Larvae gregarious in silk webs on hawthorn & blackthorn |
| 1634 | *Malacosoma neustria* | Lackey | 6 | T. Apple Loft | 01-Jul-95 | C | Deciduous trees |
| 1636 | *Lasiocampa trifolii* | Grass Eggar | 39 | Studland | 06-Aug-98 | Notable A | Coastal sandhills NW & SW Britain. F. larger than m. |
| 1637 | *Lasiocampa quercus* | Oak Eggar | 6 | T. Apple Loft | 08-Aug-96 | C | Trees and shrubs |

| 1638 | *Macrothylacia rubi* | Fox Moth | 6 | T. Charlie Wight | 31-May-96 | C | | Heathland |
|------|------|------|------|------|------|------|------|------|
| 1640 | *Euthrix potatoria* | Drinker | | T. Charlie Wight | 05-Aug-96 | C | | Grass, reeds, wet moorland |
| 1642 | *Gastropacha quercifolia* | Lappet | 38 | Durlston | 29-Jul-04 | C | | Resembles dried oak leaves, hence *quercifolia*. Blackthorn, hawthorn |
| 1643 | *Saturnia pavonia* | Emperor | 6 | T. Apple Loft | 02-May-99 | C | | Heathland on heathers, bramble etc. |
| 1644 | *Endromis versicolora* | Kentish Glory | 39 | Scotland | | Notable A | | Bred. Larvae on birch, not in Dorset |
| 1645 | *Falcaria lacertinaria* | Scalloped Hook-tip | 6 | T. Charlie Wight | 31-May-96 | C | | Birch |
| 1646 | *Drepana binaria* | Oak Hook-tip | 6 | T. Charlie Wight | 05-Aug-96 | C | | Oak |
| 1647 | *Drepana cultraria* | Barred Hook-tip | 7 | T. Fish Farm | 18-Aug-96 | L | | Beech |
| 1648 | *Drepana falcataria* | Pebble Hook-tip | 7 | T. Charlie Wight | 31-May-96 | C | | Birch |
| 1651 | *Cilix glaucata* | Chinese Character | 7 | T. Apple Loft | 25-Aug-95 | C | | Hawthorn |
| 1652 | *Thyatira batis* | Peach Blossom | 7 | T. Apple Loft | 28-Jun-95 | C | | Bramble |
| 1653 | *Habrosyne pyritoides* | Buff Arches | 7 | T. Apple Loft | 27-Jun-95 | C | | Bramble |
| 1654 | *Tethea ocularis octogesimea* | Figure of Eighty | 7 | T. Apple Loft | 17-Jun-96 | C | | Poplar, aspen |
| 1655 | *Tethea or* | Poplar Lutestring | 39 | Verwood | 15-Jun-03 | L | | 4 lines on inner forewing resemble strings on lute. Aspen |
| 1657 | *Ochropacha duplaris* | Common Lutestring | 7 | T. Charlie Wight | 05-Aug-96 | C | | Birch, oak, alder |
| 1658 | *Cymatophorima diluta hartwiegi* | Oak Lutestring | 39 | Puddletown | 16-Aug-03 | L | | Mature oak woodland |
| 1659 | *Achlya flavicornis* | Yellow Horned | 7 | T. Bottom Fish Farm | 24-Mar-96 | C | | Birch |
| 1660 | *Polyploca ridens* | Frosted Green | 7 | T. Eight Hatch | 29-Apr-96 | L | | Oak |
| 1661 | *Archiearis parthenias* | Orange Underwing | | T. Charlie Wight | 31-Mar-98 | L | | Birch |
| 1662 | *Archiearis notha* | Light Orange Underwing | 39 | King's Stag | 09-Apr-05 | Notable B | | Day flyer in sun even if cold. Mature aspen in south Britain |
| 1663 | *Alsophila aescularia* | March Moth | 7 | T. Apple Loft | 29-Apr-94 | C | | Various deciduous trees |
| 1665 | *Pseudoterpna pruinata* | Grass Emerald | | T. Charlie Wight | 05-Aug-96 | C | | Green, fades to brown. Gorse, broom |
| 1666 | *Geometra papilionaria* | Large Emerald | 7 | T. Apple Loft | 02-Jul-95 | C | | Usually birch |
| 1667 | *Comibaena bajularia* | Blotched Emerald | 39 | T. Apple Loft | 19-Jun-06 | L | | Stunning, south UK, oak. Larvae camouflaged with leaf pieces |
| 1669 | *Hemithea aestivaria* | Common Emerald | 7 | T. Charlie Wight | 22-Jul-96 | C | | Various deciduous trees and shrubs |
| 1670 | *Chlorissa viridata* | Small Grass Emerald | 7 | T. Apple Loft | 09-Jun-03 | Notable A | | Heathland, heather& birch |
| 1673 | *Hemistola chrysoprasaria* | Small Emerald | 8 | T. Ford | 13-Jul-97 | L | | Wild clematis (Old Man's Beard) |
| 1674 | *Jodis lactearia* | Little Emerald | | T. Ford | 14-Jul-96 | C | | Various deciduous trees |
| 1675 | *Cyclophora pendularia* | Dingy Mocha | 8 | T. Apple Loft | 28-Jul-01 | BAP, RDB3 | | Damp heathland with isolated sallows |
| 1677 | *Cyclophora albipunctata* | Birch Mocha | 8 | T. Charlie Wight | 31-May-96 | L | C | Birch |
| 1678 | *Cyclophora puppillaria* | Blair's Mocha | 40 | Somerset | 30-Oct-05 | Immigrant | | Scarce from south Europe. |
| 1680 | *Cyclophora punctaria* | Maiden's Blush | 8 | T. Apple Loft | 29-Jul-95 | L | | Oak |
| 1681 | *Cyclophora linearia* | Clay Triple-lines | 8 | T. South Lodge | 26-Jun-96 | L | C | Beech |
| 1682 | *Timandra comai* | Blood-vein | 8 | T. Apple Loft | 25-Jun-94 | C | | Dock, sorrel, knotgrass |
| 1689 | *Scopula marginepunctata* | Mullein Wave | 8 | T. Apple Loft | 23-May-95 | L | | Mugwort, yarrow, plantain |
| 1690 | *Scopula imitaria* | Small Blood-vein | 8 | T. Apple Loft | 04-Jul-95 | C | | Privet |
| 1691 | *Scopula emutaria* | Rosy Wave | 39 | Studland | 20-Jun-98 | Notable B | | Marshes on sea beet. Larvae are typical geometer 'loopers' |
| 1692 | *Scopula immutata* | Lesser Cream Wave | | T. Apple Loft | 01-Jul-97 | L | | Meadowsweet, valerian |
| 1693 | *Scopula floslactata* | Cream Wave | | T. Charlie Wight | 31-May-96 | L | | Dock, dandelion |
| 1698 | *Idaea muricata* | Purple-bordered Gold | 8 | T. Nowhere | 15-Jul-96 | Notable B | | Wet heaths. Flys at dawn. Marsh cinquefoil |
| 1701 | *Idaea sylvestraria* | Dotted-border Wave | 8 | T. Apple Loft | 12-Jul-97 | Notable B | | Damp heathland |
| 1702 | *Idaea biselata* | Small Fan-footed Wave | | T. Charlie Wight | 05-Aug-96 | C | | Woods, hedges |
| 1704 | *Idaea dilutaria* | Silky Wave | 39 | Wales | | BAP/RDB2 | | Bred. Limestone cliffs on common rock rose. Not in Dorset |
| 1705 | *Idaea fuscovenosa* | Dwarf Cream Wave | 8 | T. Scout Camp | 10-Jul-97 | L | | Woods, hedges |
| 1707 | *Idaea seriata* | Small Dusty Wave | | T. Apple Loft | 11-Jul-97 | C | | Ivy |

| 1708 | *Idaea dimidiata* | Single-dotted Wave | 8 | T. Charlie Wight | 22-Jul-96 | C | | Cow parsley |
|---|---|---|---|---|---|---|---|---|
| 1709 | *Idaea subsericeata* | Satin Wave | | T. Apple Loft | 20-Jun-95 | C | | Plantain |
| 1711 | *Idaea trigeminata* | Treble Brown Spot | | T. Apple Loft | 28-Jun-95 | L | | Hedges, wood edges. Ivy |
| 1713 | *Idaea aversata* | Riband Wave | 8 | T. Charlie Wight | 22-Jul-96 | C | | Low plants, bedstraw, chickweed etc. |
| 1714 | *Idaea degeneraria* | Portland Ribbon Wave | 39 | Portland | 04-Jul-97 | RDB3 | | Portland only. Food unknown |
| 1715 | *Idaea straminata* | Plain Wave | 8 | T. Charlie Wights | 22-Jul-96 | L | | Heathland |
| 1716 | *Rhodometra sacraria* | Vestal | 8 | T. Apple Loft | 03-Aug-94 | Immigrant | | Bright pink forms when breeds here |
| 1719 | *Orthonama vittata* | Oblique Carpet | 8 | T. Apple Loft | 11-Jun-97 | L | | Bedstraw in marshes |
| 1720 | *Orthonama obstipata* | Gem | 9 | T. Apple Loft | 09-Jun-96 | Immigrant | | Does not breed here |
| 1722 | *Xanthorhoe designata* | Flame Carpet | 9 | T. Ford | 07-Jun-96 | C | | Damp woodland |
| 1724 | *Xanthorhoe spadiceuaria* | Red Twin-spot Carpet | 9 | T. Apple Loft | 24-Apr-95 | C | | Bedstraw |
| 1727 | *Xanthorhoe montanata* | Silver-ground Carpet | | T. Apple Loft | 03-Jun-95 | C | | Bedstraw, damp woods, commons etc. |
| 1728 | *Xanthorhoe fluctuata* | Garden Carpet | 9 | T. Apple Loft | 25-May-96 | C | | Garlic mustard |
| 1731 | *Scotopteryx bipunctaria cretata* | Chalk Carpet | 40 | Portland | 04-Jul-97 | BAP/Ntbl. B | | Limestone cliffs, chalk downs. Bird's foot trefoil & clover |
| 1732 | *Scotopteryx chenopodiata* | Shaded Broad-bar | 9 | T. Apple Loft | 28-Jul-95 | C | | Vetch, clover |
| 1734 | *Scotopteryx luridata plumbaria* | July Belle | 9 | T. Barrow Gallop | 24-Jun-98 | C | | Heathland gorse |
| 1738 | *Epirrhoe alternata* | Common Carpet | | T. Apple Loft | 27-Jun-95 | C | | Bedstraw |
| 1739 | *Epirrhoe rivata* | Wood Carpet | 39 | Durlston | 29-Jul-04 | L | | Chalk downs, sea cliffs, wood margins on bedstraw |
| 1740 | *Epirrhoe galiata* | Galium Carpet | 40 | Portland | 04-Jul-97 | L | | Chalk downs, sea cliffs on bedstraw |
| 1742 | *Camptogramma bilineata* | Yellow Shell | 9 | T. Apple Loft | 14-Jul-97 | C | | Dock. Flies at dusk |
| 1745 | *Larentia clavaria* | Mallow | | T. Apple Loft | 12-Oct-93 | C | | Eats mallow! |
| 1746 | *Anticlea badiata* | Shoulder Stripe | 9 | T. Apple Loft | 13-Apr-95 | C | | Wild rose |
| 1747 | *Anticlea derivata* | Streamer | 9 | T. Apple Loft | 25-Apr-95 | C | | Wild rose |
| 1748 | *Mesoleuca albicillata* | Beautiful Carpet | | T. Apple Loft | 23-Jul-95 | C | L | Woodland brambles |
| 1749 | *Pelurga comitata* | Dark Spinach | | T. Apple Loft, indoors | 15-Aug-04 | C | | Goosefoot, orache |
| 1751 | *Lampropteryx otregiata* | Devon Carpet | | T. Dynamite | 06-Aug-03 | Notable B | | Poole basin only. Bedstraw in damp woods |
| 1752 | *Cosmorhoe ocellata* | Purple Bar | 9 | T. Apple Loft | 01-Jul-94 | C | | Bedstraws |
| 1754 | *Eulithis prunata* | Phoenix | 9 | T. Ford | 13-Jul-97 | C | | Gardens on blackcurrant |
| 1755 | *Eulithis testata* | Chevron | 9 | T. Nowhere | 12-Sep-96 | C | | Heathland sallow and birch |
| 1758 | *Eulithis pyraliata* | Barred Straw | | T. Apple Loft | 02-Jul-94 | C | | Bedstraw |
| 1759 | *Ecliptopera silaceata* | Small Phoenix | | T. Apple Loft | 16-May-94 | C | | Willowherbs |
| 1760 | *Chloroclysta siterata* | Red-green Carpet | 9 | T. Apple Loft | 05-Nov-93 | C | | Oak |
| 1764 | *Chloroclysta truncata* | Common Marbled Carpet | 10 | T. Apple Loft | 28-Sep-93 | C | | Various plants, shrubs, trees |
| 1765 | *Cidaria fulvata* | Barred Yellow | 10 | T. Apple Loft | 01-Aug-96 | C | | Wild rose |
| 1766 | *Plemyria rubiginata* | Blue-bordered Carpet | 10 | T. Buddens | 10-Jul-98 | C | L | Alder, blackthorn |
| 1767 | *Thera firmata* | Pine Carpet | 10 | T. Charlie Wight | 09-Oct-96 | C | | Pines |
| 1768 | *Thera obeliscata* | Grey Pine Carpet | 10 | T. Charlie Wight | 31-May-96 | C | | Pines |
| 1769 | *Thera britannica* | Spruce Carpet | 10 | T. Apple Loft | 11-Oct-96 | C | | Norway spruce, double brooded |
| 1771.1 | *Thera cupressata* | Cypress Carpet | 10 | T. Apple Loft | 28-Oct-02 | L | | Recent immigrant on cypress, now well established |
| 1773 | *Electrophaes corylata* | Broken-barred Carpet | 10 | T. Ford | 15-May-97 | C | L | Woodland birch, oak etc. |
| 1775 | *Colostygia multistrigaria* | Mottled Grey | 40 | Durlston | 13-Feb-04 | C | | Well distributed but local, springtime, on bedstraw |
| 1776 | *Colostygia pectinaria* | Green Carpet | 10 | T. Apple Loft | 03-Jun-95 | C | | Bedstraws |
| 1777 | *Hydriomena furcata* | July Highflyer | 10 | T. Charlie Wight | 05-Aug-96 | C | | Abundant on sallow, hazel, heather |
| 1778 | *Hydriomena impluviata* | May Highflyer | 10 | T.Apple Loft | 21-Jun-96 | C | | Alder |
| 1779 | *Hydriomena ruberata* | Ruddy Highflyer | 10 | T.Apple Loft | 17-Jun-03 | L | | Woods, heath on sallows |

| 1781 | *Horisme vitalbata* | Small Waved Umber | 39 | Seacombe | 17-Aug-02 | C | | Double-brooded on wild clematis (old man's beard/traveller's joy) |
|------|---------------------|-------------------|----|----------|-----------|---|---|---|
| 1784 | *Melanthia procellata* | Pretty Chalk Carpet | 40 | Langton M. | 16-Jun-01 | C | | Larvae on wild clematis (old man's beard/traveller's joy) |
| 1789 | *Rheumaptera undulata* | Scallop Shell | | T.Apple Loft | 08-Aug-96 | C | | Woods, marsh on sallows |
| 1792 | *Philereme transversata* | Dark Umber | 10 | T. Dynamite | 16-Jul-04 | L | | Downland, buckthorn |
| 1793 | *Euphyia biangulata* | Cloaked Carpet | 9 | T. Scout Camp | 03-Aug-95 | Notable B | | Woodland edges, chickweed |
| 1794 | *Euphyia unangulata* | Sharp-angled Carpet | 10 | T. Apple Loft | 01-Aug-96 | L | | Woodland edges, chickweed |
| 1795 | *Epirrita dilutata* | November Moth | 10 | T. Apple Loft | 18-Oct-97 | C | | Many trees & shrubs |
| 1797 | *Epirrita autumnata* | Autumnal | 10 | T. Apple Loft | 18-Oct-95 | C | L | Birch woods, heath |
| 1799 | *Operophtera brumata* | Winter Moth | 10 | T. Apple Loft | 30-Nov-93 | C | | Abundant, most trees & shrubs |
| 1800 | *Operophtera fagata* | Northern Winter Moth | 10 | T. Apple Loft | 07-Dec-96 | C | L | Birch, apple |
| 1802 | *Perizoma affinitata* | Rivulet | 11 | T. Apple Loft | 22-Jun-02 | C | | Red campion |
| 1803 | *Perizoma alchemillata* | Small Rivulet | 11 | T. Apple Loft | 02-Jul-95 | C | | Hemp-nettle |
| 1806 | *Perizoma blandiata* | Pretty Pinion | 40 | Harris | 06-Jul-04 | Notable B | | Scotland & Outer Hebrides. Flowers and seeds of eyebright |
| 1807 | *Perizoma albulata albulata* | Grass Rivulet | 40 | Cashmoor | 17-Jun-05 | L | | Open dry limestone grassy areas. Larvae on yellow rattle seeds |
| 1808 | *Perizoma flavofasciata* | Sandy Carpet | 11 | T. Apple Loft | 11-Jul-95 | C | | Campions |
| 1809 | *Perizoma didymata* | Twin-spot Carpet | 40 | Puddletown | 13-Jun-03 | C | | Bilberry, sallow, heather etc. |
| 1811 | *Eupithecia tenuiata* | Slender Pug | | T. Apple Loft | 22-Jul-97 | C | | Sallow catkins in damp woods |
| 1813 | *Eupithecia haworthiata* | Haworth's Pug | 40 | Ballard Down | 19-Jul-03 | L | | Orange abdomen. On traveller's joy/old man's beard |
| 1814 | *Eupithecia plumbeolata* | Lead-coloured Pug | 40 | Verwood | 15-Jun-03 | Notable B | | Common cow wheat |
| 1816 | *Eupithecia linariata* | Toadflax Pug | | T. Ford | 23-Aug-97 | C | L | Downs on toadflax |
| 1817 | *Eupithecia pulchellata* | Foxglove Pug | 11 | T. Charlie Wight | 22-Jul-96 | C | | General on foxglove |
| 1818 | *Eupithecia irriguata* | Marbled Pug | 40 | Powerstock | 10-Apr-04 | Notable B | | On old oaks |
| 1819 | *Eupithecia exiguata* | Mottled Pug | 40 | Holt | 29-May-03 | C | | Widespread, on blackthorn, hawthorn etc. |
| 1823 | *Eupithecia venosata* | Netted Pug | 40 | Durlston | 27-May-05 | L | | Widespread, limestone cliffs, downs. Bladder or sea campion |
| 1825 | *Eupithecia centaureata* | Lime-speck Pug | 11 | T. Apple Loft | 18-Jun-94 | C | | Many herbaceous plants |
| 1830 | *Eupithecia absinthiata* | Wormwood Pug | | T. Scout Camp | 10-Jul-97 | C | | On yarrow, ragwort etc |
| 1831 | *Eupithecia goossensiata* | Ling Pug | | T. Barrow Gallop | 22-Apr-98 | L | | Heather |
| 1832 | *Eupithecia assimilata* | Currant Pug | | T. Apple Loft | 09-May-02 | C | | Wild hop, currant |
| 1834 | *Eupethecia vulgata* | Common Pug | 11 | T. Apple Loft | 26-May-97 | C | | Sallow, hawthorn, ragwort etc |
| 1835 | *Eupithecia tripunctaria* | White-spotted Pug | 11 | T. Apple Loft | 16-May-00 | L | | Woods, rivers on elder, parsnip seeds |
| 1837 | *Eupithecia subfuscata* | Grey Pug | 11 | T. Ford | 05-Apr-97 | C | | Various |
| 1838 | *Eupithecia icterata* | Tawny-speckled Pug | 11 | T. Apple Loft | 19-Jul-97 | C | | Yarrow |
| 1842 | *Eupithecia simpliciata* | Plain Pug | 11 | T. Apple Loft | 06-Aug-05 | L | DR | Salt-marshes, goosefoot seeds |
| 1844 | *Eupithecia indigata* | Ochreous Pug | | T. Duck Ponds | 02-Jun-96 | C | | Scots pine shoots |
| 1846 | *Eupithecia nanata angusta* | Narrow-winged Pug | 11 | T. Charlie Wight | 31-May-96 | C | | Heather |
| 1852 | *Eupithecia abbreviata* | Brindled Pug | 11 | T. Apple Loft | 22-Apr-95 | C | | Oak |
| 1853 | *Eupithecia dodoneata* | Oak-tree Pug | 11 | T. Badgers Alley | 12-May-98 | C | | Oak, hawthorn |
| 1856 | *Eupithecia lariciata* | Larch Pug | | T. Apple Loft | 17-May-97 | C | | Larch |
| 1857 | *Eupithecia tantillaria* | Dwarf Pug | 11 | T. Badgers Alley | 19-May-01 | C | | Coniferous woodland on Norway spruce & Douglas fir |
| 1858 | *Chloroclystis v-ata* | V-Pug | 11 | T. Charlie Wight | 22-Jul-96 | C | | Varied plants |
| 1860 | *Pasiphila rectangulata* | Green Pug | 11 | T. Margaret's Garden | 07-Jul-96 | C | | Apples, blackthorn |
| 1861 | *Pasiphila debiliata* | Bilberry Pug | | T. Apple Loft | 04-Jul-01 | Notable B | | 1st Dorset Record. Bilberry |
| 1862 | *Gymnoscelis rufifasciata* | Double-striped Pug | 11 | T. Charlie Wight | 22-Jul-96 | C | | Many plants |
| 1863 | *Anticollix sparsata* | Dentated Pug | 11 | T. Apple Loft | 07-Jul-00 | Notable A | | East Dorset on yellow loosestrife in shade |
| 1864 | *Chesias legatella* | Streak | 12 | T. Apple Loft | 22-Oct-96 | C | L | Broom |

| 1867 | Aplocera plagiata | Treble Bar | 12 | T. Ford | 03-Jun-97 | C | | St John's-wort |
|------|-------------------|------------|----|---------|-----------|---|---|----------------|
| 1874 | Euchoeca nebulata | Dingy Shell | 12 | T. Ford | 07-Jun-96 | L | | Alder groves |
| 1875 | Asthena albulata | Small White Wave | | T. Buddens | 14-Jun-97 | C | | Hazel, birch |
| 1876 | Hydrelia flammeolaria | Small Yellow Wave | 12 | T. Brick Kiln | 31-May-97 | C | | Maple or alder |
| 1879 | Lobophora halterata | Seraphim | 12 | T. Ford | 07-Jun-96 | L | | Poplar, aspen |
| 1881 | Trichopteryx carpinata | Early Tooth-striped | 12 | T. Badgers Alley | 29-Mar-98 | C | | Honeysuckle, birch, sallow, alder |
| 1882 | Pterapherapteryx sexalata | Small Seraphim | | T. Charlie Wight | 05-Aug-96 | L | | Damp woods on sallow |
| 1883 | Acasis viretata | Yellow-barred Brindle | 12 | T. Apple Loft | 02-May-99 | L | | Holly, ivy, privet, dogwood, guelder rose |
| 1884 | Abraxas grossulariata | Magpie Moth | 12 | T. Ford | 13-Jul-97 | C | | Sundry inc hawthorn hazel etc. |
| 1885 | Abraxas sylvata | Clouded Magpie | 40 | Suffolk | 20-Jun-03 | L | | Widespread, woods & parks. On wych or English elm |
| 1887 | Lomaspilis marginata | Clouded Border | 12 | T. Charlie Wight | 31-May-96 | C | | Sallow, aspen, also poplar or hazel |
| 1888 | Ligdia adustata | Scorched Carpet | 12 | T. Apple Loft | 27-Apr-96 | L | | Spindle |
| 1889 | Macaria notata | Peacock | 12 | T. Apple Loft | 06-Aug-01 | L | | Birch |
| 1890 | Macaria alternaria | Sharp-angled Peacock | | T. Apple Loft | 10-Jul-94 | L | C | Sallow, blackthorn, alder |
| 1893 | Macaria liturata | Tawny-barred Angle | 12 | T. Ford | 07-Jun-96 | C | | Scots and Norway pine |
| 1894 | Chiasmia clathrata | Latticed Heath | 12 | T. Nowhere | days | C | | Day/night flyer. Lucerne, clovers & trefoils |
| 1901 | Cepphis advenaria | Little Thorn | 12 | T. Scout Camp | 04-Jul-95 | Notable B | | Bilberry, bramble, wild rose or dogwood |
| 1902 | Petrophora chlorosata | Brown Silver-line | 13 | T. Charlie Wight | 31-May-96 | C | | Bracken |
| 1903 | Plagodis pulveraria | Barred Umber | 40 | Holt | 29-May-03 | L | | Widespread but local on oak, birch, sallow |
| 1904 | Plagodis dolabraria | Scorched Wing | 13 | T. Apple Loft | 02-Jun-95 | L | | Oak, birch, sallow etc. |
| 1905 | Pachycnemia hippocastanaria | Horse Chestnut | 13 | T. Charlie Wight | 05-Aug-96 | Notable B | C | Heathland on heather, occurs almost all year round |
| 1906 | Opisthograptis luteolata | Brimstone | 13 | T. Charlie Wight | 09-Oct-96 | C | | Hawthorn, backthorn, rowan |
| 1907 | Epione repandaria | Bordered Beauty | 13 | T. Eight Hatch | 29-Sep-96 | C | L | Damp woodland on sallow |
| 1909 | Pseudopanthera macularia | Speckled Yellow | 13 | T. Charlie Wight | 30-May-96 | C | | Flies in sunshine. Wood sage |
| 1910 | Apeira syringaria | Lilac Beauty | | T. Apple Loft | 28-Jun-95 | L | | Woods. Honeysuckle, privet, ash |
| 1912 | Ennomos quercinaria | August Thorn | | T. Ford | 19-Aug-96 | L | | Woodland. Oak, beech, birch etc. |
| 1913 | Ennomos alniaria | Canary-shouldered Thorn | 13 | T. Apple Loft | 14-Sep-93 | C | | Woodland. Birch, sallow, alder, lime |
| 1914 | Ennomos fuscantaria | Dusky Thorn | 13 | T. Charlie Wight | 09-Oct-96 | C | | Ash |
| 1915 | Ennomos erosaria | September Thorn | | T. Apple Loft | 10-Sep-93 | C | | Oak, lime, birch |
| 1917 | Selenia dentaria | Early Thorn | 13 | T. Apple Loft | 24-Apr-94 | C | | Sundry trees, shrubs. Double brooded |
| 1919 | Selenia tetralunaria | Purple Thorn | 13 | T. Apple Loft | 28-Apr-94 | C | | Sundry trees, shrubs. Double brooded |
| 1920 | Odontopera bidentata | Scalloped Hazel | 13 | T. Charlie Wight | 31-May-96 | C | | Sundry deciduous and coniferous trees |
| 1921 | Crocallis elinguaria | Scalloped Oak | | T. Charlie Wight | 22-Jul-96 | C | | Most deciduous trees, shrubs |
| 1922 | Ourapteryx sambucaria | Swallow-tailed Moth | 14 | T. Apple Loft | 28-Jun-94 | C | | Ivy and other trees, shrubs |
| 1923 | Colotois pennaria | Feathered Thorn | 14 | T. Apple Loft | 09-Nov-93 | C | | Various trees, shrubs. Autumn flyer |
| 1925 | Apocheima hispidaria | Small Brindled Beauty | | T. Apple Loft | 24-Mar-95 | L | | Old oak woodland |
| 1926 | Apocheima pilosaria | Pale Brindled Beauty | 14 | T. Apple Loft | 16-Mar-96 | C | | Wingless female. Various trees, shrubs |
| 1927 | Lycia hirtaria | Brindled Beauty | 14 | T. Ford | 21-Apr-96 | C | | Birch, alder,hawthorn, beech etc. |
| 1930 | Biston strataria | Oak Beauty | 14 | T. Apple Loft | 24-Mar-95 | C | | Oak, elm, hazel etc |
| 1931 | Biston betularia | Peppered Moth | 15 | T. Charlie Wight | 31-May-96 | C | | Variable colour forms. Various trees & plants |
| 1932 | Agriopis leucophaearia | Spring Usher | 14 | T. Apple Loft | 02-Mar-97 | C | | Oak |
| 1933 | Agriopis aurantiaria | Scarce Umber | 15 | T. Apple Loft | 09-Dec-96 | C | | Autumn. Birch, oak, hazel etc. |
| 1934 | Agriopis marginaria | Dotted Border | 15 | T. Apple Loft | 23-Mar-96 | C | | Wingless female. Various trees, shrubs |
| 1935 | Erannis defoliaria | Mottled Umber | 15 | T. Apple Loft | 31-Dec-95 | C | | Many trees & shrubs, inc pyracantha |
| 1936 | Menophra abruptaria | Waved Umber | 15 | T. Apple Loft | 30-Apr-95 | C | | Privet, lilac, ash, birch |

| | | | | | | | | |
|---|---|---|---|---|---|---|---|---|
| 1937 | *Peribatodes rhomboidaria* | Willow Beauty | 16 | T. Apple Loft | 23-Jun-96 | C | | Privet, yew, plum, ivy, traveller's joy etc. |
| 1938 | *Selidosema brunnearia* | Bordered Grey | 15 | T. Apple Loft | 11-Aug-03 | Notable A | | Heather. Will fly by day and night |
| 1939 | *Cleora cinctaria* | Ringed Carpet | | T. Badgers Alley | 28-Apr-98 | Notable A | | Lightly wooded heathland only. Birch, heather |
| 1941 | *Alcis repandata* | Mottled Beauty | 16 | T. Charlie Wight | 22-Jul-96 | C | | Birch, bramble, many others |
| 1944 | *Serraca punctinalis* | Pale Oak Beauty | 16 | T. Brick Kiln | 31-May-97 | C | | Woodland on oak, birch |
| 1945 | *Cleorodes lichenaria* | Brussels Lace | 16 | T. Apple Loft | 03-Jul-95 | L | C | Lichens on oak or blackthorn |
| 1947 | *Ectropis bistortata* | Engrailed | 16 | T. Ford | 07-Apr-96 | C | | Double-brooded. Various trees, shrubs |
| 1948 | *Ectropis crepuscularia* | Small Engrailed | | T. Apple Loft | 22-Apr-95 | L | | Birch, larch, sallow etc. |
| 1949 | *Paradarisa consonaria* | Square Spot | 16 | T. Duck Ponds | 02-Jun-96 | L | | Woods on oak, birch, pine, beech, yew |
| 1951 | *Aethalura punctulata* | Grey Birch | | T. Ford | 07-Jun-96 | C | | Birch, alder |
| 1952 | *Ematurga atomaria* | Common Heath | | T. Charlie Wight | 18-Jul-96 | C | | Day flying on heaths. Heather, clover |
| 1954 | *Bupalus piniaria* | Bordered White | 16 | T. Apple Loft | 19-Jul-95 | C | | Pines |
| 1955 | *Cabera pusaria* | Common White Wave | 16 | T. Charlie Wight | 22-Jul-96 | C | | Birch, sallow, alder etc. |
| 1956 | *Cabera exanthemata* | Common Wave | | T. Charlie Wight | 05-Aug-96 | C | | Damp woods. Sallow, aspen |
| 1957 | *Lomographa bimaculata* | White-pinion Spotted | 41 | Puddletown | 13-Jun-03 | C | | Hawthorn, blackthorn in woods |
| 1958 | *Lomographa temerata* | Clouded Silver | 16 | T. Charlie Wight | 05-Aug-96 | C | | Hawthorn, blackthorn, plum, cherry |
| 1960 | *Theria primaria* | Early Moth | 16 | T. Broad Meadow | 15-Feb-97 | C | | Hawthorn, blackthorn. Flightless females |
| 1961 | *Campaea margaritata* | Light Emerald | | T. Charlie Wight | 31-May-96 | C | | Oak, birch, beech, elm etc. |
| 1962 | *Hylaea fasciaria* | Barred Red | 16 | T. Apple Loft | 06-Jun-95 | C | | Conifers |
| 1964 | *Charissa obscurata* | Annulet | 17 | T. Hyde Heath | 29-Jul-97 | L | | Light forms on chalk, dark on heathland. Heather, sundry plants |
| 1968 | *Aspitates ochrearia* | Yellow Belle | 17 | T. North Drive Pond | 07-Aug-01 | L | | Near coasts. Wild carrot, buck's-horn plantain etc. |
| 1969 | *Dyscia fagaria* | Grey Scalloped Bar | | T. Hyde Heath | 30-May-97 | L | | Heather, heath, cross-leaved heath |
| 1970 | *Perconia strigillaria* | Grass Wave | 17 | T. Apple Loft | 20-Jun-02 | L | | Heathland. Heather, heath, broom, gorse |
| 1972 | *Agrius convolvuli* | Convolvulus Hawk-moth | 18 | T. Duck Ponds | 26-Sep-98 | Immigrant | | Field bindweed *Convolvulus* |
| 1973 | *Acherontia atropos* | Death's Head Hawk-moth | 36 | | | Immigrant | | Bred. Huge larvae feed on potatoes |
| 1976 | *Sphinx ligustri* | Privet Hawk-moth | 17 | T. Apple Loft | 30-May-94 | C | | Privet, lilac, ash, holly |
| 1978 | *Hyloicus pinastri* | Pine Hawk-moth | 18 | T. Charlie Wights | 05-Aug-96 | L | C | Scot's pine, Norway spruce |
| 1979 | *Mimas tiliae* | Lime Hawk-moth | 18 | T. Apple Loft | 28-Jun-95 | C | | Lime, elm, alder, birch |
| 1980 | *Smerinthus ocellata* | Eyed Hawk-moth | 18 | T. Apple Loft | 24-May-95 | C | | Willow, sallow, aspen, apple |
| 1981 | *Laothoe populi* | Poplar Hawk-moth | 19 | T. Charlie Wight | 31-May-96 | C | | Popla, aspen, sallow, willow |
| 1983 | *Hemaris fuciformis* | Broad-bordered Bee Hawk-moth | | T. Charlie Wight | 18-Jul-96 | Notable B | | Woodland rides on honeysuckle. Day flyer |
| 1984 | *Macroglossum stellatarum* | Humming-bird Hawk-moth | | T. Apple Loft | 19-Aug-03 | Immigrant | | Bedstraws |
| 1990 | *Hyles lineata livornica* | Striped Hawk-moth | 41 | Wisconsin USA | 11-Sep-04 | Immigrant | | Regular migrant from Europe |
| 1991 | *Deilephila elpenor* | Elephant Hawk-moth | 19 | T. Charlie Wight | 22-Jul-96 | C | | Willow herbs, bedstraw |
| 1992 | *Deilephila porcellus* | Small Elephant Hawk-moth | 19 | T. Apple Loft | 01-Jul-96 | L | | Downs & heaths etc. Bedstraw |
| 1994 | *Phalera bucephala* | Buff-tip | 20 | T. Charlie Wight | 31-May-96 | C | | Elm, sallow, lime, oak,etc |
| 1995 | *Cerura vinula* | Puss Moth | 20 | T. Bottom Fish Farm | 25-May-96 | C | | Poplar, sallow, aspen |
| 1997 | *Furcula furcula* | Sallow Kitten | 20 | T. Charlie Wight | 31-May-96 | C | | Double-brooded.Sallow, aspen, poplar |
| 1999 | *Stauropus fagi* | Lobster Moth | 20 | T. Apple Loft | 02-Jul-94 | C | | Mature woods,. Oak, hazel, birch, beech |
| 2000 | *Notodonta dromedarius* | Iron Prominent | 20 | T. Charlie Wight | 05-Aug-96 | C | | Double brooded. Birch, alder, hazel, oak |
| 2003 | *Eligmodonta ziczac* | Pebble Prominent | 20 | T. Apple Loft | 07-May-94 | C | | Double brooded. Sallow, willow, aspen, poplar |
| 2005 | *Peridea anceps* | Great Prominent | 20 | T. Apple Loft | 06-May-94 | L | C | Oak |
| 2006 | *Pheosia gnoma* | Lesser Swallow Prominent | 20 | T. Charlie Wight | 31-May-96 | C | | Double brooded. Birch |
| 2007 | *Pheosia tremula* | Swallow Prominent | | T. Apple Loft | 21-May-95 | C | | Double brooded. Willow, sallow, poplar, aspen |
| 2008 | *Ptilodon capucina* | Coxcomb Prominent | 21 | T. Apple Loft | 02-Jun-94 | C | | Double brooded. Hawthorn, birch, alder, oak etc. |

| 2010 | *Odontosia carmelita* | Scarce Prominent | 21 | T. Badgers Alley | 28-Apr-98 | L | | Mature woods of birch |
|------|----------------------|------------------|----|------------------|-----------|---|---|----------------------|
| 2011 | *Pterostoma palpina* | Pale Prominent | 21 | T. Apple Loft | 15-May-94 | C | | Poplar, aspen, sallow |
| 2014 | *Drymonia dodonaea* | Marbled Brown | 21 | T. Charlie Wight | 31-May-96 | C | | Oak |
| 2015 | *Drymonia ruficornis* | Lunar Marbled Brown | 21 | T. Apple Loft | 01-May-94 | L | | Oak |
| 2017 | *Clostera pigra* | Small Chocolate-tip | 21 | T. Apple Loft | 24-Apr-03 | Notable B | | Heathland. Creeping & eared willow, sallow |
| 2019 | *Clostera curtula* | Chocolate-tip | 21 | T. Apple Loft | 27-May-95 | L | | Willow, poplar, aspen |
| 2026 | *Orgyia antiqua* | Vapourer | 21 | T. Apple Loft | 21-Sep-97 | C | | Most deciduous trees, shrubs. Wingless female |
| 2027 | *Dicallomera fascelina* | Dark Tussock | 21 | T. Barrow Gallop | 09-Jul-99 | L | | Heathland. Heather, sallow, broom etc. |
| 2028 | *Calliteara pudibunda* | Pale Tussock | 22 | T. Charlie Wight | 31-May-96 | C | | Birch, oak, elm, lime etc. |
| 2029 | *Euproctis chrysorrhoea* | Brown-tail | | T. Apple Loft | 02-Jul-95 | L | C | Hawthorn, blackthorn etc. |
| 2030 | *Euproctis similis* | Yellow-tail | | T. Charlie Wight | 22-Jul-96 | C | | Hawthorn, blackthorn etc. |
| 2031 | *Leucoma salicis* | White Satin | 22 | T. Charlie Wight | 22-Jul-96 | L | | Sallow, willow, poplar |
| 2033 | *Lymantria monacha* | Black Arches | 21 | T. Scout Camp | 18-Jun-98 | L | C | Oak woodland |
| 2034 | *Lymantria dispar* | Gypsy | 41 | Durlston | 05-Aug-03 | Immigrant | | Extinct resident, now migrant males only. Pest of decidous trees |
| 2035 | *Thumatha senex* | Round-winged Muslin | 22 | T. Brick Kiln | 16-Jul-96 | L | | Marshy places. Lichens, moss |
| 2037 | *Miltochrista miniata* | Rosy Footman | 22 | T. Charlie Wight | 22-Jul-96 | L | C | Woodland lichens |
| 2039 | *Atolmis rubricollis* | Red-necked Footman | 22 | T. Buddens | 14-Jun-97 | L | | Annual fluctuations. Lichens on tree trunks |
| 2040 | *Cybosia mesomella* | Four-dotted Footman | 22 | T. Charlie Wight | 22-Jul-96 | L | | Heaths, open woods on lichens, algae |
| 2043 | *Eilema sororcula* | Orange Footman | 22 | T. Apple Loft | 02-May-99 | L | | Increasing. Lichens on mature oaks |
| 2044 | *Eilema griseola* | Dingy Footman | 22 | T. Apple Loft | 20-Aug-93 | C | | Damp woods. Lichens |
| 2047 | *Eilema complana* | Scarce Footman | 22 | T. Charlie Wight | 05-Aug-96 | L | | Heaths and woods on lichens |
| 2049 | *Eilema deplana* | Buff Footman | 22 | T. Apple Loft | 01-Jul-95 | L | C | Lichens & algae on yew, hawthorn etc. |
| 2050 | *Eilema lurideola* | Common Footman | | T. Charlie Wight | 22-Jul-96 | C | | Lichens on trees, fences, rocks |
| 2053 | *Coscinia cribraria* | Speckled Footman | 22 | T. Trigon | 29-Jul-98 | BAP/RDB1 | | Heaths.One of Britain's rarest moths. Curtis's hair grass |
| 2057 | *Arctia caja* | Garden Tiger | 23 | T. Apple Loft | 16-Jul-94 | C | | Sundry garden plants |
| 2058 | *Arctia villica britannica* | Cream-spot Tiger | 22 | T. Apple Loft | 05-Jun-98 | L | | Coastal. Various herbaceous plants |
| 2059 | *Diacrisia sannio* | Clouded Buff | 22 | T. Apple Loft | 03-Jul-95 | L | | Heathland, on heather & haeth |
| 2060 | *Spilosoma lubricipeda* | White Ermine | 23 | T. Apple Loft | 15-May-95 | C | | Various wild & garden plants |
| 2061 | *Spilosoma luteum* | Buff Ermine | 23 | T. Charlie Wight | 22-Jul-96 | C | | Various wild & garden plants |
| 2063 | *Diaphora mendica* | Muslin Moth | 23 | T. Charlie Wight | 31-May-96 | C | | Dock, chickweed etc. Shows sexual dimorphism, m. dark grey,  f. white |
| 2064 | *Phragmatobia fuliginosa* | Ruby Tiger | | T. Charlie Wight | 05-Aug-96 | C | | Dock, dandelion, etc. |
| 2068 | *Callimorpha dominula* | Scarlet Tiger | 23 | T. Trinity Dam | 10-May-98 | L | | Larva on cat! Comfrey, nettle, bramble etc. |
| 2069 | *Tyria jacobaeae* | Cinnabar | 23 | T. Charlie Wight | 31-May-96 | C | | Ragwort, groundsel |
| 2075 | *Meganola strigula* | Small Black Arches | 41 | Verwood | 15-Jun-03 | Notable A | | South UK, mature oak woods |
| 2076 | *Meganola albula* | Kent Black Arches | 24 | T. Cowleaze | 20-Jul-97 | Notable B | | Bramble, strawberry, raspberry, dewberry |
| 2077 | *Nola cucullatella* | Short-cloaked | | T. Apple Loft | 21-Jul-96 | C | | Hawthorn, blackthorn, apple, plum |
| 2078 | *Nola confusalis* | Least Black Arches | 24 | T. South Lodge | 09-Jun-96 | L | | Lichens on oak, beech |
| 2081 | *Euxoa tritici* | White-line Dart | 24 | T. Apple Loft | 17-Aug-96 | C | | Small herbaceous plants |
| 2084 | *Agrotis cinerea* | Light Feathered Rustic | | T. Apple Loft | 04-May-97 | Notable B | | Coastal. Thyme |
| 2085 | *Agrotis vestigialis* | Archer's Dart | 24 | T. Apple Loft | 30-Jul-95 | L | | Coastal heaths and grassland |
| 2087 | *Agrotis segetum* | Turnip | 24 | T. Charlie Wight | 09-Oct-96 | C | | Roots of root vegetables etc. |
| 2088 | *Agrotis clavis* | Heart and Club | | T. Apple Loft | 07-Jul-94 | C | | Roots & leaves of dock, clover, knotgrass etc |
| 2089 | *Agrotis exclamationis* | Heart and Dart | | T. Charlie Wight | 31-May-96 | C | | Various plants |
| 2090 | *Agrotis trux lunigera* | Crescent Dart | 24 | T. Brick Kiln | 11-Jul-99 | L | | Coastal. Various low plants |
| 2091 | *Agrotis ipsilon* | Dark Sword-grass | 24 | T. Apple Loft | 05-Aug-04 | Immigrant | | Roots & leaves of various plants |

| 2092 | *Agrotis puta* | Shuttle Shaped Dart | 24 | T. Charlie Wight | 05-Aug-96 | C | | Dock, dandelion etc. |
|------|----------------|---------------------|----|-----------------|-----------|---|---|----------------------|
| 2093 | *Agrotis ripae* | Sand Dart | 41 | Studland | 20-Jun-98 | Notable B | | Maritime sand dunes. Prickly saltwort etc. Night feeder, hides in sand |
| 2098 | *Axylia putris* | Flame | 24 | T. Apple Loft | 20-Jun-94 | C | | Dock, dandelion etc. |
| 2102 | *Ochropleura plecta* | Flame Shoulder | 24 | T. Charlie Wight | 05-Aug-96 | C | | Plantain, dock, groundsel etc. |
| 2104 | *Standfussiana lucernea* | Northern Rustic | 41 | Ballard Down | 19-Jul-03 | L | | South and west UK coastal cliffs, quarries. On grasses or stonecrop |
| 2107 | *Noctua pronuba* | Large Yellow Underwing | 24 | T. Apple Loft | 02-Jul-94 | C | | Eats anything, anytime! |
| 2108 | *Noctua orbona* | Lunar Yellow Underwing | 41 | Suffolk | 11-Jun-05 | BAP/Ntbl B | | Scarce, now mainly in East Anglia. Various grasses |
| 2109 | *Noctua comes* | Lesser Yellow Underwing | 25 | T. Charlie Wight | 05-Aug-96 | C | | Trees, shrubs, plants |
| 2110 | *Noctua fimbriata* | Broad-bordered Yellow U'wing | 24 | T. Apple Loft | 12-Jul-94 | C | | Blackthorn, sallow, dock etc. |
| 2111 | *Noctua janthina* | Less. Broad-bord Yell. U'wing | | T. Apple Loft | 27-Jul-94 | C | | Blackthorn, sallow, dock etc. |
| 2112 | *Noctua interjecta caliginosa* | Least Yellow Underwing | | T. Apple Loft | 03-Aug-94 | C | | Grasses, herbs |
| 2117 | *Paradiarsia glareosa* | Autumnal Rustic | 24 | T. Apple Loft | 10-Sep-96 | C | | Heaths, woods. Various plants, grasses |
| 2118 | *Lycophotia porphyrea* | True Lover's Knot | | T. Charlie Wight | 31-May-96 | C | | Heathland heathers. Favourite nightjar food! |
| 2119 | *Peridroma saucia* | Pearly Underwing | 25 | T. Apple Loft | 11-Mar-96 | Immigrant | | Herbaceous plants |
| 2120 | *Diarsia mendica mendica* | Ingrailed Clay | 25 | T. Charlie Wight | 31-May-96 | C | | Various forms. Bramble, sallow, hawthorn etc. |
| 2122 | *Diarsia brunnea* | Purple Clay | 25 | T. Apple Loft | 22-Jun-02 | C | | Sallow, birch, dock etc in woodland |
| 2123 | *Diarsia rubi* | Small Square-spot | | T. Charlie Wight | 22-Jul-96 | C | | Various herbaceous plants |
| 2126 | *Xestia c-nigrum* | Setaceous Hebrew Character | 25 | T. Charlie Wight | 09-Oct-96 | C | | Various herbaceous plants |
| 2128 | *Xestia triangulum* | Double Square-spot | 25 | T. Charlie Wight | 22-Jul-96 | C | | Blackthorn, hawthorn, bramble etc. |
| 2130 | *Xestia baja* | Dotted Clay | 32 | T. Apple Loft | 21-Jun-03 | C | | Caterpillars hibernate, feed at night in spring on bog myrtle, bramble etc. |
| 2132 | *Xestia castanea* | Neglected Rustic | 25 | T. Nowhere | 12-Sep-96 | L | | Heathlands. Heather, erica sp. |
| 2133 | *Xestia sexstrigata* | Six-striped Rustic | 25 | T. Bottom Fish Farm | 18-Aug-96 | C | | Various herbaceous plants |
| 2134 | *Xestia xanthographa* | Square-spot Rustic | 25 | T. Apple Loft | 28-Jul-95 | C | | Grasses, low plants |
| 2135 | *Xestia agathina* | Heath Rustic | 25 | T. Nowhere | 12-Sep-96 | L | | Heather |
| 2136 | *Naenia typica* | Gothic | | T. Apple Loft | 06-Jul-96 | L | | Various, inc willowherb, dock, dandelion etc. |
| 2138 | *Anaplectoides prasina* | Green Arches | 26 | T. Buddens | 14-Jun-97 | C | | Honeysuckle |
| 2139 | *Cerastis rubricosa* | Red Chestnut | 26 | T. Apple Loft | 30-Apr-94 | C | | Various herbaceous plants |
| 2142 | *Anarta myrtilli* | Beautiful Yellow Underwing | 26 | T. Ford | 14-Jul-96 | C | L | Heather. Day/night flyer |
| 2145 | *Discestra trifolii* | Nutmeg | | T. Apple Loft | 18-Jul-95 | C | | Goosefoot, orache |
| 2147 | *Hada plebeja* | Shears | 26 | T. Charlie Wight | 31-May-96 | C | | Hawk's-beard, hawkweed, dandelion |
| 2148 | *Polia bombycina* | Pale Shining Brown | 41 | Salisbury | | BAP/Ntbl B | | On calcareous soils, life history and food imperfectly known. |
| 2150 | *Polia nebulosa* | Grey Arches | 26 | T. Apple Loft | 28-Jun-95 | C | | Various plants and shrubs |
| 2152 | *Sideridis albicolon* | White Colon | 41 | Suffolk | 11-Jun-05 | Notable B | | Coastal, dry, sandy habitats |
| 2154 | *Mamestra brassicae* | Cabbage | 26 | T. Apple Loft | 06-Aug-95 | C | | Various inc cabbages & brassicas |
| 2155 | *Melanchra persicariae* | Dot Moth | | T. Apple Loft | 27-Jun-95 | C | | Wild & garden plants, shrubs |
| 2156 | *Lacanobia contigua* | Beautiful Brocade | 26 | T. Trigon Heath | 30-May-97 | L | | Various, inc bracken, heather, sallow |
| 2157 | *Lacanobia w-latinum* | Light Brocade | 26 | T. Apple Loft | 03-Jun-95 | L | | Various inc broom, dyer's greenweed |
| 2158 | *Lacanobia thalassina* | Pale-shouldered Brocade | 26 | T. Apple Loft | 06-Jul-96 | C | | Oak, honeysuckle, hawthorn, broom etc. |
| 2159 | *Lacanobia suasa* | Dog's Tooth | 26 | T. Apple Loft | 09-Jun-97 | L | | Salt marshes, dock, plantain etc. |
| 2160 | *Lacanobia oleracea* | Bright-line Brown-eye | | T. Charlie Wight | 22-Jul-96 | C | | Various plants & shrubs |
| 2163 | *Ceramica pisi* | Broom | 26 | T. Apple Loft | 01-Jul-95 | C | L | Broom, bracken, birch etc. |
| 2164 | *Hecatera bicolorata* | Broad-barred White | | T. Barrow Gallop | 24-Jun-98 | C | | Hawkweed & hawk's-beard buds, flowers |
| 2166 | *Hadena rivularis* | Campion | 26 | T. South Lodge | 08-Jun-96 | C | | Seeds of campions |
| 2167 | *Hadena perplexa* | Tawny Shears | | T. Barrow Gallop | 25-Jun-98 | C | | Campions |
| 2172 | *Hadena albimacula* | White Spot | 41 | Ballard Down | 30-Jun-00 | BAP/RDB2 | | Coastal limestone cliffs, shingle beaches. On Nottingham catchfly |

| 2173 | *Hadena bicruris* | Lychnis | 27 | T. Charlie Wight | 22-Jul-96 | C | | Campions & lychnis |
|---|---|---|---|---|---|---|---|---|
| 2175 | *Eriopygodes imbecilla* | Silurian | 41 | Wales | | RDB2 | | Bred. Discovered 1972. Herbaceous plants, grasses. Not in Dorset |
| 2176 | *Cerapteryx graminis* | Antler | 42 | Perthshire | 05-Jul-03 | C | | Grassy down and moorland |
| 2177 | *Tholera cespitis* | Hedge Rustic | 27 | T. Apple Loft | 25-Aug-95 | C | | Mat-grass, other grasses |
| 2178 | *Tholera decimalis* | Feathered Gothic | 27 | T. Apple Loft | 05-Sep-97 | C | | Grasses |
| 2179 | *Panolis flammea* | Pine Beauty | 27 | T. Apple Loft | 04-Apr-95 | C | | Pines |
| 2181 | *Egira conspicillaris* | Silver Cloud | 42 | Gloucester | 07-May-03 | Notable A | | Severn valley only. Food unknown |
| 2182 | *Orthosia cruda* | Small Quaker | 27 | T. Apple Loft | 06-Apr-95 | C | | Oak, sallow etc. |
| 2186 | *Orthosia gracilis* | Powdered Quaker | 27 | T. Apple Loft | 30-Apr-94 | C | | Bog myrtle, sallow, willow, loosestrifes etc. |
| 2187 | *Orthosia cerasi* | Common Quaker | 27 | T. Apple Loft | 26-Mar-96 | C | | Oak, sallow etc. |
| 2188 | *Orthosia incerta* | Clouded Drab | 27 | T. Ford | 07-Apr-96 | C | | Oak, sallow etc. |
| 2189 | *Orthosia munda* | Twin-spotted Quaker | 27 | T. Ford | 01-Apr-96 | C | | Sallow, elm, oak etc. |
| 2190 | *Orthosia gothica* | Hebrew Character | 27 | T. Charlie Wight | 31-May-96 | C | | Many trees, plants |
| 2191 | *Mythimna turca* | Double Line | 42 | Melbury | 13-Jul-02 | BAP/Ntbl B | | South UK in mature woodland. Wood-rush, wood meadow-grass etc. |
| 2192 | *Mythimna conigera* | Brown-line Bright Eye | 27 | T. Dynamite | 27-Jul-01 | C | L | Grasses i.e. cock's-foot, common couch etc. |
| 2193 | *Mythimna ferrago* | Clay | | T. Charlie Wight | 22-Jul-96 | C | | Grasses, low plants |
| 2194 | *Mythimna albipuncta* | White-point | 27 | T. Apple Loft | 20-Aug-97 | Immigrant | | Grasses. Breeds on south coast |
| 2195 | *Mythimna vitellina* | Delicate | 28 | T. Apple Loft | 19-Sep-97 | Immigrant | | Grasses |
| 2196 | *Mythimna pudorina* | Striped Wainscot | | T. Charlie Wight | 22-Jul-96 | L | | Wet heath. Common reed, purple moor grass etc. |
| 2197 | *Mythimna straminea* | Southern Wainscot | 28 | T. Nowhere | 13-Aug-96 | L | | Marsh. Common reed, canary grass |
| 2198 | *Mythimna impura* | Smoky Wainscot | | T. Apple Loft | 04-Jul-95 | C | | Various grasses |
| 2199 | *Mythimna pallens* | Common Wainscot | 28 | T. Apple Loft | 29-May-95 | C | | Various grasses |
| 2200 | *Mythimna favicolor* | Mathew's Wainscot | 28 | T. Margaret's Garden | 06-Jul-96 | Notable B | | Poole basin only. Common saltmarsh grass |
| 2201 | *Mythimna litoralis* | Shore Wainscot | 42 | Studland | 20-Jun-98 | Notable B | | Coastal sand dunes only, on marram |
| 2202 | *Mythimna l-album* | L-album Wainscot | 28 | T. Apple Loft | 08-Oct-96 | Notable B | | South coast. Grasses |
| 2203 | *Mythimna unipuncta* | White-speck | | T. Apple Loft | 19-Jul-95 | Immigrant | | Couch grass |
| 2204 | *Mythimna obsoleta* | Obscure Wainscot | | T. Brick Kiln | 31-May-97 | L | | Common reed |
| 2205 | *Mythimna comma* | Shoulder-striped Wainscot | 28 | T. Apple Loft | 07-Jul-94 | C | | Cock's-foot etc. |
| 2208 | *Mythimna loreyi* | Cosmopolitan | | T. Apple Loft | 12-Oct-96 | Immigrant | | Cock's-foot etc. |
| 2209 | *Mythimna flammea* | Flame Wainscot | 42 | Arne | 18-May-02 | Notable A | | Poole basin and East Anglia only. On common reed, *Phragmites* |
| 2214 | *Cucullia chamomillae* | Chamomile Shark | 28 | T. Lower Hyde Heath | 12-Jun-97 | L | | Mayweeds & chamomiles |
| 2216 | *Cucullia umbratica* | Shark | | T. Apple Loft | 09-Jul-94 | C | | Sow-thistle, wild lettuce |
| 2221 | *Cucullia verbasci* | Mullein | 28 | T. Apple Loft | 30-Apr-95 | C | | Mulleins (*Verbascum*) and *Buddleia* |
| 2225 | *Brachylomia viminalis* | Minor Shoulder-knot | | T. Apple Loft | 21-Jul-96 | C | | Willow, sallow |
| 2226 | *Leucochlaena oditis* | Beautiful Gothic | 42 | St Aldhelm's Head | 06-Oct-96 | RDB3 | | Common on Purbeck coast, scarce otherwise. Grasses on limestone |
| 2229 | *Dasypolia templi* | Brindled Ochre | 42 | Durlston | 24-Oct-96 | L | | Rocky coasts, larvae in roots of hogweed |
| 2230 | *Aporophyla australis pascuea* | Feathered Brindle | 42 | St Aldhelm's Head | 06-Oct-96 | Notable B | | South UK coasts. Sand, shingle, cliffs on sea campion and grasses |
| 2231 | *Aporophyla lutulenta* | Deep-brown Dart | 29 | T. Charlie Wight | 09-Oct-96 | C | | Southern UK. Tufted hair-grass etc. |
| 2232 | *Aporophyla nigra* | Black Rustic | 29 | T. Charlie Wight | 09-Oct-96 | C | | Western UK. Dock, heather, grasses |
| 2235 | *Lithophane semibrunnea* | Tawny Pinion | 28 | T. Apple Loft | 18-Oct-00 | L | | Annual fluctuations. Open woodland. Ash |
| 2236 | *Lithophane hepatica* | Pale Pinion | 28 | T. Apple Loft | 15-Apr-96 | L | | Hibernates. Sallow etc. |
| 2237 | *Lithophane ornitopus lactipennis* | Grey Shoulder-knot | 28 | T. Apple Loft | 13-Oct-95 | C | | Hibernates. Oak woods. |
| 2240 | *Lithophane leautieri hesperica* | Blair's Shoulder-knot | 28 | T. Charlie Wight | 09-Oct-96 | C | | Monterey cypress & *leylandii* |
| 2243 | *Xylocampa areola* | Early Grey | 28 | T. Apple Loft | 24-Mar-95 | C | | Variable colour. Honeysuckle |
| 2245 | *Allophyes oxyacanthae* | Green-brindled Crescent | 29 | T. Apple Loft | 22-Oct-96 | C | | Woods, hedges. Blackthorn, hawthorn |

| 2246.1 | *Dryobota labecula* | Oak Rustic | 43 | Durlston | 26-Oct-05 | Immigrant | | New for Britain. Feeds on evergreen oak, *Quercus ilex* |
|---|---|---|---|---|---|---|---|---|
| 2247 | *Dichonia aprilina* | Merveille du Jour | 29 | T. Apple Loft | 07-Nov-93 | C | | Stunning moth! Isolated oaks |
| 2248 | *Dryobotodes eremita* | Brindled Green | 29 | T. Apple Loft | 23-Sep-96 | C | | Oak, hawthorn |
| 2252 | *Polymixis flavicincta* | Large Ranunculus | 42 | Norfolk | 08-Sep-96 | L | | Widespread, often coastal. Ragwort, red valerian & sundry |
| 2255 | *Polymixis lichenea* | Feathered Ranunculus | 29 | T. Duck Ponds | 27-Sep-97 | L | | Mostly coastal. Various plants inc stonecrop |
| 2256 | *Eupsilia transversa* | Satellite | | T. Apple Loft | 08-Apr-95 | C | | On oak, birch, elm etc but also cannibalistic |
| 2258 | *Conistra vaccinii* | Chestnut | | T. Apple Loft | 12-Oct-95 | C | | Woodland through winter. Oak, birch, elm etc. |
| 2260 | *Conistra rubiginea* | Dotted Chestnut | 29 | T. Apple Loft | 23-Apr-96 | Notable B | | SW England. Hibernates. Food unknown |
| 2262 | *Agrochola circellaris* | Brick | 29 | T. Charlie Wight | 09-Oct-96 | C | | Variable. On wych elm flowers & seeds |
| 2263 | *Agrochola lota* | Red-line Quaker | 29 | T. Apple Loft | 13-Oct-93 | C | | Woods, heaths etc. Sallow, willow |
| 2264 | *Agrochola macilenta* | Yellow-line Quaker | 29 | T. Apple Loft | 04-Nov-93 | C | | Beech, poplar catkins, Hawthorn, oak |
| 2264.1 | *Agrochola haematidea* | Southern Chestnut | 42 | Parley | 28-Oct-03 | RDB2 | | Found 1990. Dry, acid heaths Dorset, Hants, Sussex. Bell heather |
| 2267 | *Agrochola lychnidis* | Beaded Chestnut | 30 | T. Charlie Wight | 09-Oct-96 | C | | Variable. Grasses |
| 2268 | *Parastichtis suspecta* | Suspected | 30 | T. Charlie Wight | 22-Jul-96 | L | | Woods, heaths. Birch, sallow |
| 2269 | *Atethmia centrago* | Centre-barred Sallow | 30 | T. Eight Hatch | 29-Sep-96 | C | | Ash buds & leaves |
| 2270 | *Omphaloscelis lunosa* | Lunar Underwing | 30 | T. Apple Loft | 17-Sep-95 | C | | Blades, stems, roots of various grasses |
| 2271 | *Xanthia citrago* | Orange Sallow | | T. Roundabout Pond | 24-Sep-98 | C | L | Lime |
| 2272 | *Xanthia aurago* | Barred Sallow | 30 | T. Apple Loft | 07-Oct-02 | C | L | Beech, maple |
| 2273 | *Xanthia togata* | Pink-barred Sallow | 30 | T. Eight Hatch | 29-Sep-96 | C | | Sallows, low plants |
| 2274 | *Xanthia icteritia* | Sallow | 30 | T. Eight Hatch | 29-Sep-96 | C | | Sallows, low plants |
| 2278 | *Acronicta megacephala* | Poplar Grey | | T. Apple Loft | 08-Aug-96 | C | | Poplar, aspen, willow |
| 2279 | *Acronicta aceris* | Sycamore | | T. Apple Loft | 05-Jul-95 | L | | Horse chestnut, sycamore, field maple, oak |
| 2280 | *Acronicta leporina* | Miller | 30 | T. Apple Loft | 28-Jun-95 | C | | Woods, heaths, commons. Birch, oak, sallow etc. |
| 2281 | *Acronicta alni* | Alder | 30 | T. Buddens | 11-Jun-96 | L | | Woodland, birch, oak, elm, hawthorn etc. |
| 2284 | *Acronicta psi* | Grey Dagger | 31 | T. Apple Loft | 25-Jun-95 | C | | Birch, oak, lime, rowan, hawthorn etc. |
| 2289 | *Acronicta rumicis* | Knot Grass | 31 | T. Apple Loft | 22-Aug-93 | C | | Double brooded. Plantain, dock, thistle, hop etc. |
| 2291 | *Craniophora ligustri* | Coronet | | T. Charlie Wight | 05-Aug-96 | L | | Woods etc. Ash, privet |
| 2293 | *Cryphia domestica* | Marbled Beauty | 31 | T. Charlie Wight | 05-Aug-96 | C | | Urban areas & coasts. Lichens on walls, rocks |
| 2295 | *Cryphia muralis* | Marbled Green | | T. Apple Loft | 05-Aug-97 | L | | South UK coastal. Lichens on rocks |
| 2297 | *Amphipyra pyramidea* | Copper Underwing | 31 | T. Apple Loft | 15-Aug-93 | C | | Oak, honeysuckle, ash, privet etc. |
| 2299 | *Amphipyra tragopogonis* | Mouse Moth | 31 | T. Apple Loft | 09-Aug-93 | C | | Various leaves & flowers, inc. mugwort, fennel |
| 2300 | *Motmo maura* | Old Lady | 31 | T. Apple Loft | 17-Jul-04 | L | | Blackthorn, hawthorn, elm, sallow, birch |
| 2301 | *Dypterygia scabriuscula* | Bird's Wing | 31 | T. Apple Loft | 16-Jul-95 | L | | Dock, sorrel etc. in central south UK |
| 2302 | *Rusina ferruginea* | Brown Rustic | 31 | T. Charlie Wight | 31-May-96 | C | | Dock, plantain, groundsel |
| 2303 | *Thalpophila matura* | Straw Underwing | 31 | T. Apple Loft | 23-Aug-05 | C | | Grasses |
| 2305 | *Euplexia lucipara* | Small Angle Shades | 32 | T. Apple Loft | 01-Jul-94 | C | | Bracken, ferns, ivy, willowherb etc. |
| 2306 | *Phlogophora meticulosa* | Angle Shades | 32 | T. Charlie Wight | 31-May-96 | C | | Feeds on anything, anytime! |
| 2311 | *Ipimorpha retusa* | Double Kidney | | T. Bottom Fish Farm | 01-Aug-97 | L | | Damp & marshy woods. Sallow, willow |
| 2312 | *Ipimorpha subtusa* | Olive | | T. Barrow Gallop | 18-Jul-99 | L | | Spun leaves aspen, poplar |
| 2318 | *Cosmia trapezina* | Dun-bar | 32 | T. Charlie Wight | 05-Aug-96 | C | | Various trees, shrubs & cannibalistic |
| 2319 | *Cosmia pyralina* | Lunar-spotted Pinion | 32 | T. Apple Loft | 11-Jul-95 | L | | Elm, wych elm |
| 2321 | *Apamea monoglypha* | Dark Arches | | T. Charlie Wight | 31-May-96 | C | | Sundry grasses |
| 2322 | *Apamea lithoxylea* | Light Arches | | T. Charlie Wight | 22-Jul-96 | C | | Sundry grasses |
| 2326 | *Apamea crenata* | Clouded-bordered Brindle | 32 | T. Apple Loft | 06-Jun-94 | C | | Woods, downs, moors. Cock's-foot etc. |
| 2327 | *Apamea epomidion* | Clouded Brindle | | T. Buddens | 14-Jun-97 | C | | Various grasses |

| | | | | | | | | |
|---|---|---|---|---|---|---|---|---|
| 2330 | *Apamea remissa* | Dusky Brocade | 32 | T. Charlie Wight | 31-May-96 | C | | Various grasses |
| 2331 | *Apamea unanimis* | Small Clouded Brindle | | T. Buddens | 11-Jun-96 | C | | Various grasses in damp woods |
| 2334 | *Apamea sordens* | Rustic Shoulder-knot | 43 | St Aldhelm's | 15-May-00 | C | | Grassy habitats on cock's foot , grasses |
| 2335 | *Apamea scolopacina* | Slender Brindle | 32 | T. Apple Loft | 17-Aug-96 | L | | Woodland grasses |
| 2336 | *Apamea ophiogramma* | Double Lobed | 32 | T. Ford | 13-Jul-97 | L | | Damp woods, fens. Canary & sweet grass |
| 2337 | *Oligia strigilis* | Marbled Minor | | T. Apple Loft | 22-Jun-94 | C | | Cock's-foot, reed canary-grass etc. |
| 2338 | *Oligia versicolor* | Rufous Minor | | T. Apple Loft | 01-Jul-95 | L | | Probably grasses |
| 2339 | *Oligia latruncula* | Tawny Marbled Minor | 32 | T. Apple Loft | 11-Jun-94 | C | | Various grasses |
| 2340 | *Oligia fasciuncula* | Middle-barred Minor | 33 | T. Ford | 19-Jun-96 | C | | Various grasses, damp habitats |
| 2342 | *Mesoligia literosa* | Rosy Minor | 33 | T. Apple Loft | 17-Aug-96 | C | | Various grasses inc marram, sheep's-fescue |
| 2343 | *Mesapamea secalis* | Common Rustic | | T. Apple Loft | 01-Jul-94 | C | | Cock's-foot, tall fescue etc. |
| 2343.1 | *Mesapamea didyma* | Lesser Common Rustic | | T. Charlie Wight | 22-Jul-96 | C | | Cock's-foot, tall fescue etc. |
| 2345 | *Photedes minima* | Small Dotted Buff | 33 | T. Ford | 19-Jun-96 | C | | Damp meadows and woods. Tufted hair-grass |
| 2350 | *Photedes pygmina* | Small Wainscot | 33 | T. Apple Loft | 09-Sep-95 | C | | Damp meadows and woods. Sedges & rushes |
| 2352 | *Eremobia ochroleuca* | Dusky Sallow | 33 | T. Apple Loft | 19-Jul-95 | C | | SE UK. Cock's-foot, couch, oat-grass etc. |
| 2353 | *Luperina testacea* | Flounced Rustic | | T. Apple Loft | 06-Sep-93 | C | | Variable, grassland. Grass roots |
| 2358 | *Amphipoea fucosa paludis* | Saltern Ear | 33 | T. Apple Loft | 11-Aug-97 | L | | Salt-marshes. Annual meadow-grass |
| 2360 | *Amphipoea oculea* | Ear | 33 | T. Apple Loft | 17-Aug-96 | C | | Damp habitats. Tufted hair-grass etc. |
| 2361 | *Hydraecia micacea* | Rosy Rustic | 33 | T. Charlie Wight | 05-Aug-96 | C | | Dock, burdock, plantain etc. |
| 2364 | *Gortyna flavago* | Frosted Orange | 33 | T. Apple Loft | 25-Sep-93 | C | | Burdock, foxglove, thistles |
| 2365 | *Gortyna borelii lunata* | Fisher's Estuarine | 43 | Essex | 25-Sep-04 | RDB2 | | Very local, on hog's fennel. Protected species |
| 2368 | *Celaena leucostigma* | Crescent | | T. Apple Loft | 02-Aug-97 | L | | Marshes. Yellow iris, great fen-sedge |
| 2369 | *Nonagria typhae* | Bulrush Wainscot | | T. Roundabout Pond | 03-Aug-96 | C | | Stems of common reedmace (bulrush) |
| 2370 | *Archanara geminipuncta* | Twin-spotted Wainscot | 33 | T. Roundabout Pond | 07-Aug-96 | L | | Stems of common reed |
| 2371 | *Archanara dissoluta* | Brown-veined Wainscot | 33 | T. Brick Kiln | 25-Aug-01 | L | | Stems of common reed |
| 2373 | *Archanara sparganii* | Webb's Wainscot | | T. Apple Loft | 14-Jul-95 | Notable B | | Isolated colonies. Stems of bulrush, iris etc. |
| 2375 | *Rhizedra lutosa* | Large Wainscot | 33 | T. Apple Loft | 21-Oct-96 | C | | Variable size. Wanders. Common reed |
| 2376 | *Sedina buettneri* | Blair's Wainscot | 33 | T. Duck Ponds | 06-Oct-97 | RDB1 | DR | Formerly I O Wight only. *Carex acutiformis* |
| 2377 | *Arenostola phragmitidis* | Fen Wainscot | | T. Charlie Wight | 05-Aug-96 | L | | SE England. Common reed stems |
| 2379 | *Coenobia rufa* | Small Rufous | | T. Bottom Fish Farm | 01-Aug-97 | L | | Stems of jointed & soft rush |
| 2380 | *Charanyca trigrammica* | Treble Lines | 34 | T. Apple Loft | 21-May-95 | C | | Variable. Plantain, knapweed, dwarf thistle |
| 2381 | *Hoplodrina alsines* | Uncertain | | T. Ford | 14-Jul-96 | C | | Dock, plantain, dandelion |
| 2382 | *Hoplodrina blanda* | Rustic | | T. Charlie Wight | 31-May-96 | C | | Dock, plantain etc. |
| 2384 | *Hoplodrina ambigua* | Vine's Rustic | | T. Charlie Wight | 09-Oct-96 | L | | SE UK. Chickweed, dandelion, knotgrass |
| 2385 | *Spodoptera exigua* | Small Mottled Willow | 34 | T. Charlie Wight | 31-May-96 | Immigrant | | Low plants, if at all |
| 2387 | *Caradrina morpheus* | Mottled Rustic | | T. Apple Loft | 18-Jun-96 | C | | Nettle, dandelion etc. |
| 2389 | *Caradrina clavipalpis* | Pale Mottled Willow | | T. Apple Loft | 02-May-95 | C | | Cereal ears, plantain seeds etc. |
| 2391 | *Chilodes maritimus* | Silky Wainscot | 34 | T. Margarets Garden | 06-Jul-96 | L | | Coastal reed-beds. Forages in old stems |
| 2394 | *Stilbia anomala* | Anomalous | 34 | T. Apple Loft | 22-Sep-98 | L | | Heath & moor. Wavy hair-grass |
| 2396 | *Elaphria venustula* | Rosy Marbled | | T. Ford Plantation | 17-Jun-98 | Notable B | | New as Dorset resident. Possibly tormentil |
| 2397 | *Panemeria tenebrata* | Small Yellow Underwing | 43 | Shapwick | 14-May-03 | L | | Active in sunshine. Feeds on common mouse-ear |
| 2399 | *Pyrrhia umbra* | Bordered Sallow | 43 | Portland | 04-Jul-97 | L | | Mainly coastal, on flowers and seeds of restharrow |
| 2400 | *Helicoverpa armigera* | Scarce Bordered Straw | | T. Charlie Wight | 09-Oct-96 | Immigrant | | Seen most years |
| 2403 | *Heliothis peltigera* | Bordered Straw | 34 | T. Apple Loft | 27-Aug-03 | Immigrant | | Common, sometimes breeds, restharrow etc. |
| 2410 | *Lithacodia pygarga* | Marbled White Spot | | T. Apple Loft | 18-Jun-95 | C | | Purple moor-grass |

| 2412 | *Eustrotia uncula* | Silver Hook | 34 | T. Apple Loft | 28-Jul-00 | L | | Boggy heaths & marsh. Sedge, tufted hair-grass |
|---|---|---|---|---|---|---|---|---|
| 2418 | *Earias clorana* | Cream-bordered Green Pea | 34 | T. Ford | 08-Jun-97 | Notable B | | Damp woods etc. Willows, osier |
| 2421 | *Bena bicolorana* | Scarce Silver-lines | | T. Apple Loft | 28-Jul-95 | L | | Oak |
| 2422 | *Pseudoips fagana* | Green Silver-lines | 34 | T. Apple Loft | 20-Jun-94 | C | | Woods. Oak, birch, beech, hazel etc. |
| 2423 | *Nycteola revayana* | Oak Nycteoline | 34 | T. Apple Loft | 22-Jul-96 | L | | Extremely variable, see pictures. Hibernates. On oak |
| 2425 | *Colocasia coryli* | Nut-tree Tussock | 34 | T. Charlie Wight | 31-May-96 | C | | Double-brooded. Hazel, beech, hornbeam etc. |
| 2432 | *Trichoplusia ni* | Ni | 35 | T. Apple Loft | 10-Sep-96 | Immigrant | | Scarce |
| 2434 | *Diachrysia chrysitis* | Burnished Brass | 34 | T. Apple Loft | 26-Jun-94 | C | | Double-brooded. Nettle |
| 2439 | *Plusia festucae* | Gold Spot | 35 | T. Apple Loft | 24-Aug-01 | C | | Double-brooded. River banks. Yellow iris |
| 2441 | *Autographa gamma* | Silver Y | 34 | T. Charlie Wight | 31-May-96 | Immigrant | | Common and breeds in summer on anything |
| 2442 | *Autographa pulchrina* | Beautiful Golden Y | 35 | T. Hyde Heath | 18-Jun-97 | C | | Woodland. Honeysuckle, dead-nettle etc. |
| 2443 | *Autographa jota* | Plain Golden Y | 43 | Stockley | 18-Jun-97 | C | | Feeds on nettle, honeysuckle etc. |
| 2449 | *Abrostola triplasia* | Dark Spectacle | 35 | T. Ford | 19-Jun-96 | C | L | Western Britain. Hop & nettle |
| 2450 | *Abrostola tripartita* | Spectacle | 35 | T. Apple Loft | 22-Apr-94 | C | | Woods, gardens etc. Nettle |
| 2452 | *Catacola nupta* | Red Underwing | 35 | T. Apple Loft | 30-Oct-96 | C | L | Willow, poplar, aspen |
| 2462 | *Callistege mi* | Mother Shipton | 43 | Shapwick | 14-May-03 | C | | Active in sunshine. Feeds on clover |
| 2463 | *Euclidea glyphica* | Burnet Companion | 43 | Shapwick | 14-May-03 | C | | Active in sunshine. Feeds on clovers and trefoils |
| 2466 | *Lygephila pastinum* | Blackneck | | T. Buddens | 14-Jun-97 | L | | Damp woods & meadows. Tufted vetch |
| 2469 | *Scoliopteryx libatrix* | Herald | 35 | T. Apple Loft | 11-Jul-95 | C | | Hibernates in sheds, barns. Sallow, willow etc. |
| 2470 | *Phytometra viridaria* | Small Purple-barred | 35 | T. Apple Loft | 19-Jul-95 | L | | Day/night flyer. Woods, heaths. Milkwort |
| 2473 | *Laspeyria flexula* | Beautiful Hook-tip | 35 | T. Apple Loft | 25-Jun-94 | L | | Lichens on hawthorn, larch, yew, apple |
| 2474 | *Rivula sericealis* | Straw Dot | 35 | T. Apple Loft | 15-Aug-93 | C | | Damp woods & heaths. Torgrass, false brome |
| 2477 | *Hypena proboscidalis* | Snout | 35 | T. Apple Loft | 25-Jun-94 | C | | Double-brooded. Nettle |
| 2484 | *Schrankia costaestrigalis* | Pinion-streaked Snout | 35 | T. Nowhere | 13-Aug-96 | L | | Boggy heaths & woods. Food unknown |
| 2485 | *Hypenodes humidalis* | Marsh Oblique-barred | | T. Margaret's Garden | 18-Aug-96 | Notable B | | Boggy heaths. Food unknown |
| 2489 | *Herminia tarsipennalis* | Fan-foot | | T. Apple Loft | 05-Jul-95 | C | | Withered leaves of beech, bramble, oak etc. |
| 2492 | *Herminia grisealis* | Small Fan-foot | 35 | T. Apple Loft | 16-Jul-95 | C | | Withered & fresh leaves of oak & alder |

# ENGLISH NAMES. Alphabetical with Log No. and Photo Plate No.

| Log No. | English Name | Plate |
|---|---|---|
| 2281 | Alder | 30 |
| 2306 | Angle Shades | 32 |
| 1964 | Annulet | 17 |
| 2394 | Anomalous m. | 34 |
| 2176 | Antler | 42 |
| 2085 | Archer's Dart | 24 |
| 1912 | August Thorn | |
| 1797 | Autumnal | 10 |
| 2117 | Autumnal Rustic | 24 |
| 970 | Barred Fruit-tree Tortrix, micro | |
| 1647 | Barred Hook-tip | 7 |
| 1962 | Barred Red | 16 |
| 2272 | Barred Sallow | 30 |
| 1758 | Barred Straw | |
| 1903 | Barred Umber | 40 |
| 1765 | Barred Yellow | 10 |
| 2267 | Beaded Chestnut | 30 |
| 2156 | Beautiful Brocade | 26 |
| 1748 | Beautiful Carpet | |
| 1350 | Beautiful China Mark, pyrale | 4 |
| 2442 | Beautiful Golden Y | 35 |
| 2226 | Beautiful Gothic | 42 |
| 2473 | Beautiful Hook-tip | 35 |
| 2142 | Beautiful Yellow Underwing | 26 |
| 1428 | Bee moth, pyrale | 5 |
| 1861 | Bilberry Pug | |
| 1677 | Birch Mocha | 8 |
| 424 | Bird-cherry Ermine, micro | 1 |
| 2301 | Bird's Wing | 31 |
| 2033 | Black Arches | 21 |
| 2232 | Black Rustic | 29 |
| 2466 | Blackneck | |
| 1678 | Blair's Mocha | 40 |
| 2240 | Blair's Shoulder-knot | 28 |
| 2376 | Blair's Wainscot | 33 |
| 1682 | Blood-vein | 8 |
| 1667 | Blotched Emerald | 39 |
| 1766 | Blue-bordered Carpet | 10 |
| 1907 | Bordered Beauty | 13 |
| 1938 | Bordered Grey | 15 |
| 2399 | Bordered Sallow | 43 |
| 2403 | Bordered Straw | 34 |
| 1954 | Bordered White | 16 |
| 1175 | Bramble Shoot, micro | |
| 2262 | Brick | 29 |
| 2160 | Bright-line Brown-eye | |
| 1906 | Brimstone | 13 |
| 1927 | Brindled Beauty | 14 |
| 2248 | Brindled Green | 29 |
| 2229 | Brindled Ochre | 42 |
| 1852 | Brindled Pug | 11 |
| 2164 | Broad-barred White | |
| 2110 | Broad-bord. Yellow U'wing | 24 |
| 1984 | Broad-bordered Bee Hawk | |
| 1773 | Broken-barred Carpet | 10 |
| 2163 | Broom | 26 |
| 1345 | Brown China-mark, pyrale | 4 |
| 647 | Brown House Moth, micro | |
| 2302 | Brown Rustic | 31 |
| 1902 | Brown Silver-line | 13 |
| 2192 | Brown-line Bright Eye | 27 |
| 2029 | Brown-tail | |
| 2371 | Brown-veined Wainscot | 33 |
| 1945 | Brussels Lace | 16 |
| 1653 | Buff Arches | 7 |
| 2061 | Buff Ermine | 23 |
| 2049 | Buff Footman | 22 |
| 1994 | Buff-tip | 20 |
| 2369 | Bulrush Wainscot | 33 |
| 2463 | Burnet Companion | 43 |
| 2434 | Burnished Brass | 34 |
| 2154 | Cabbage | 26 |
| 2166 | Campion | 26 |
| 1913 | Canary-shouldered Thorn | 13 |
| 985 | Carnation Tortrix, micro | |
| 2269 | Centre-barred Sallow | 30 |
| 1731 | Chalk Carpet | 40 |
| 2214 | Chamomile Shark | 28 |
| 969 | Chequered Fruit-tree Tortrix, mic | |
| 2258 | Chestnut | |
| 1755 | Chevron | 9 |
| 1651 | Chinese Character | 7 |
| 2019 | Chocolate-tip | 21 |
| 2069 | Cinnabar | 23 |
| 2193 | Clay | |
| 1681 | Clay Triple-lines | 8 |
| 1793 | Cloaked Carpet | 9 |
| 1887 | Clouded Border | 12 |
| 2327 | Clouded Brindle | |
| 2059 | Clouded Buff | 22 |
| 2188 | Clouded Drab | 27 |
| 1885 | Clouded Magpie | 40 |
| 1958 | Clouded Silver | 16 |
| 2326 | Clouded-bordered Brindle | 32 |
| 1261 | Codling, micro | |
| 1738 | Common Carpet | |
| 1669 | Common Emerald | 7 |
| 2050 | Common Footman | |
| 1952 | Common Heath | |
| 1657 | Common Lutestring | 7 |
| 1764 | Common Marbled Carpet | 10 |
| 1834 | Common Pug | 11 |
| 2187 | Common Quaker | 27 |
| 2343 | Common Rustic | |
| 17 | Common Swift | 1 |
| 2199 | Common Wainscot | 28 |
| 1956 | Common Wave | |
| 1955 | Common White Wave | 16 |
| 1972 | Convolvulus Hawk | 18 |
| 2297 | Copper Underwing | 31 |
| 216 | Cork Moth, micro | |
| 2291 | Coronet | |
| 2208 | Cosmopolitan | |
| 2008 | Coxcomb Prominent | 21 |
| 1693 | Cream Wave | |
| 2418 | Cream-bord'ed Green Pea | 34 |
| 2058 | Cream-spot Tiger | 22 |
| 2368 | Crescent | |
| 2090 | Crescent Dart | 24 |
| 373 | Currant Clearwing | 1 |
| 1832 | Currant Pug | |
| 1771.1 | Cypress Carpet | 10 |
| 2321 | Dark Arches | |
| 972 | Dark Fruit-tree Tortrix, micro | |
| 2449 | Dark Spectacle | 35 |
| 1749 | Dark Spinach | |
| 2091 | Dark Sword-grass | 24 |
| 2027 | Dark Tussock | 21 |
| 1792 | Dark Umber | 10 |
| 1973 | Death's Head Hawk | 36 |
| 1631 | December Moth | 6 |
| 2231 | Deep-brown Dart | 29 |
| 2195 | Delicate | 28 |
| 1863 | Dentated Pug | 11 |
| 1751 | Devon Carpet | |
| 464 | Diamond-back Moth, micro | |
| 2044 | Dingy Footman | 22 |
| 1675 | Dingy Mocha | 8 |
| 1874 | Dingy Shell | 12 |
| 2159 | Dog's Tooth | 26 |
| 2155 | Dot Moth | |
| 1934 | Dotted Border | 15 |
| 1701 | Dotted Border Wave | 8 |
| 2260 | Dotted Chestnut | 29 |
| 2130 | Dotted Clay | 32 |
| 2311 | Double Kidney | |
| 2191 | Double Line | 42 |
| 2336 | Double Lobed | 32 |
| 2128 | Double Square-spot | 25 |
| 1862 | Double-striped Pug | 11 |
| 1640 | Drinker | |
| 2318 | Dun-bar | 32 |
| 2330 | Dusky Brocade | 32 |
| 2352 | Dusky Sallow | 33 |
| 1914 | Dusky Thorn | 13 |
| 1705 | Dwarf Cream Wave | 8 |
| 1857 | Dwarf Pug | 11 |
| 2360 | Ear | 33 |
| 2243 | Early Grey | 29 |
| 1960 | Early Moth | 16 |
| 1917 | Early Thorn | 13 |
| 1881 | Early Tooth-striped | 12 |
| 1991 | Elephant Hawk | 19 |
| 1643 | Emperor | 6 |
| 1947 | Engrailed | 16 |
| 1375 | European Corn-borer, pyrale | 4 |
| 1980 | Eyed Hawk | 18 |
| 2489 | Fan-foot | |
| 2230 | Feathered Brindle | 42 |
| 2178 | Feathered Gothic | 27 |
| 2255 | Feathered Ranunculus | 29 |
| 1923 | Feathered Thorn | 14 |
| 2377 | Fen Wainscot | |
| 173 | Festoon | 1 |
| 1654 | Figure of Eighty | 7 |
| 332.1 | Firethorn Leaf Miner, micro | 1 |
| 2365 | Fisher's Estuarine | 43 |
| 2098 | Flame | 24 |
| 1722 | Flame Carpet | 9 |
| 2102 | Flame Shoulder | 24 |
| 2209 | Flame Wainscot | 42 |
| 2353 | Flounced Rustic | |
| 2040 | Four-dotted Footman | 22 |
| 1638 | Fox | 6 |
| 1817 | Foxglove Pug | 11 |
| 1660 | Frosted Green | 7 |
| 2364 | Frosted Orange | 33 |
| 1740 | Galium Carpet | 40 |
| 1728 | Garden Carpet | 9 |
| 1293 | Garden Grass-veneer, pyrale | |
| 1356 | Garden Pebble, pyrale | |
| 1048 | Garden Rose Tortrix, micro | 2 |
| 2057 | Garden Tiger | 23 |
| 1720 | Gem | 9 |
| 14 | Ghost Moth | 1 |
| 162 | Goat | 37 |
| 2439 | Gold Spot | 35 |
| 16 | Gold Swift | 1 |
| 1413 | Gold Triangle, pyrale | 4 |
| 2136 | Gothic | |
| 1636 | Grass Eggar | 39 |
| 1665 | Grass Emerald | |
| 1807 | Grass Rivulet | 40 |
| 1970 | Grass Wave | 17 |
| 2005 | Great Prominent | 20 |
| 2138 | Green Arches | 26 |
| 1776 | Green Carpet | 10 |
| 1033 | Green Oak Tortrix, micro | 2 |
| 1860 | Green Pug | 11 |
| 2422 | Green Silver-lines | 34 |
| 2245 | Green-brindled Crescent | 29 |
| 2150 | Grey Arches | 26 |
| 1951 | Grey Birch | |
| 2284 | Grey Dagger | 31 |
| 1768 | Grey Pine Carpet | 10 |
| 1837 | Grey Pug | 11 |
| 1969 | Grey Scalloped Bar | |
| 2237 | Grey Shoulder-knot | 28 |
| 1020 | Grey Tortrix, micro | |
| 2034 | Gypsy | 41 |
| 1813 | Haworth's Pug | 40 |
| 2088 | Heart and Club | |
| 2089 | Heart and Dart | |
| 2135 | Heath Rustic | 25 |
| 2190 | Hebrew Character | 27 |
| 2177 | Hedge Rustic | 27 |
| 2469 | Herald | 35 |
| 453 | Honeysuckle Moth, micro | |
| 1905 | Horse Chestnut | 13 |
| 1984 | Humming-bird Hawk | |
| 2120 | Ingrailed Clay | 25 |
| 2000 | Iron Prominent | 20 |
| 1734 | July Belle | 9 |
| 1777 | July Highflyer | 10 |
| 2076 | Kent Black Arches | 24 |
| 1644 | Kentish Glory | 39 |
| 2289 | Knot Grass | 31 |
| 1634 | Lackey | 6 |
| 2202 | L-album Wainscot | 28 |
| 1642 | Lappet | 38 |
| 1856 | Larch Pug | |
| 1666 | Large Emerald | 7 |
| 977 | Large Fruit-tree Tortrix, micro | 2 |
| 2252 | Large Ranunculus | 42 |
| 2375 | Large Wainscot | 33 |
| 2107 | Large Yellow Underwing | 24 |
| 1894 | Latticed Heath | 12 |
| 1814 | Lead-coloured Pug | 40 |
| 2078 | Least Black Arches | 24 |
| 2112 | Least Yellow Underwing | |
| 161 | Leopard | 9 |
| 2111 | Less. Broad-bord Yell. U'wing | |
| 2343 | Lesser Common Rustic | |
| 1692 | Lesser Cream Wave | |
| 2006 | Lesser Swallow Prominent | 20 |
| 2109 | Lesser Yellow Underwing | 25 |
| 2322 | Light Arches | |
| 2157 | Light Brocade | 26 |
| 1961 | Light Emerald | |
| 2084 | Light Feathered Rustic | |
| 1662 | Light Orange Underwing | 39 |
| 1910 | Lilac Beauty | |
| 1979 | Lime Hawk | 18 |
| 1825 | Lime-speck Pug | 11 |
| 1831 | Ling Pug | |
| 1674 | Little Emerald | |
| 1901 | Little Thorn | 12 |
| 1999 | Lobster Moth | 20 |
| 2015 | Lunar Marbled Brown | 21 |
| 2270 | Lunar Underwing | 30 |
| 2108 | Lunar Yellow Underwing | 41 |
| 2319 | Lunar-spotted Pinion | 32 |
| 2173 | Lychnis | 27 |
| 1884 | Magpie Moth | 12 |
| 1680 | Maiden's Blush | 8 |
| 1745 | Mallow | |
| 2293 | Marbled Beauty | 31 |
| 2014 | Marbled Brown | 21 |
| 2295 | Marbled Green | |
| 2337 | Marbled Minor | |

# ENGLISH NAMES. Alphabetical with Log No. and Photo Plate No.

| Log No. | Name | Log No. | Name | Log No. | Name | Log No. | Name | Log No. | Name |
|---|---|---|---|---|---|---|---|---|---|
| 1818 | Marbled Pug **40** | 1652 | Peach Blossom **7** | 1997 | Sallow Kitten **20** | 1948 | Small Engrailed | 2381 | Uncertain |
| 2410 | Marbled White Spot | 1889 | Peacock **12** | 2358 | Saltern Ear **33** | 2492 | Small Fan-foot **35** | 2026 | Vapourer **21** |
| 1663 | March. **7** | 2119 | Pearly Underwing **25** | 2093 | Sand Dart **41** | 1702 | Small Fan-footed Wave | 1716 | Vestal **8** |
| 2485 | Marsh Oblique-barred | 1648 | Pebble Hook-tip **7** | 1808 | Sandy Carpet **11** | 1670 | Small Grass Emerald **7** | 2384 | Vine's Rustic |
| 2200 | Mathew's Wainscot **28** | 2003 | Pebble Prominent **20** | 2256 | Satellite | 1376 | Small Magpie, pyrale **4** | 1858 | V-Pug **11** |
| 1778 | May Highflyer **10** | 1931 | Peppered **15** | 1709 | Satin Wave | 2385 | Small Mottled Willow **34** | 1331 | Water Veneer, pyrale |
| 2247 | Merveille du Jour **29** | 1754 | Phoenix **9** | 1789 | Scallop Shell | 1759 | Small Phoenix | 1936 | Waved Umber **15** |
| 2340 | Middle-barred Minor **33** | 2179 | Pine Beauty **27** | 1920 | Scalloped Hazel **13** | 2470 | Small Purple-barred **35** | 1425 | Wax, pyrale **5** |
| 2280 | Miller **30** | 1209 | Pine Bud Moth, micro | 1645 | Scalloped Hook-tip **6** | 2182 | Small Quaker **27** | 2373 | Webb's Wainscot |
| 2225 | Minor Shoulder-knot | 1767 | Pine Carpet **10** | 1921 | Scalloped Oak | 1803 | Small Rivulet **11** | 2152 | White Colon **41** |
| 1405 | Mother of Pearl, pyrale **4** | 1978 | Pine Hawk **18** | 2400 | Scarce Bordered Straw | 2379 | Small Rufous | 2060 | White Ermine **23** |
| 2462 | Mother Shipton **43** | 2484 | Pinion-streaked Snout **35** | 2047 | Scarce Footman **22** | 1882 | Small Seraphim | 1513 | White plume **5** |
| 1941 | Mottled Beauty **16** | 2273 | Pink-barred Sallow **30** | 2010 | Scarce Prominent **21** | 2123 | Small Square-spot | 2031 | White Satin **22** |
| 1775 | Mottled Grey **40** | 2443 | Plain Golden Y **43** | 2421 | Scarce Silver-lines | 2350 | Small Wainscot **33** | 2172 | White Spot **41** |
| 1819 | Mottled Pug **40** | 1842 | Plain Pug **11** | 1933 | Scarce Umber **15** | 1781 | Small Waved Umber **39** | 2081 | White-line Dart **24** |
| 2387 | Mottled Rustic | 1715 | Plain Wave **8** | 2068 | Scarlet Tiger **23, front'pce** | 1875 | Small White Wave | 1957 | White-pinion Spotted **41** |
| 1935 | Mottled Umber **15** | 1082 | Plum Totrix, micro **2** | 1888 | Scorched Carpet **12** | 2397 | Small Yellow Underwing **43** | 2194 | White-point **27** |
| 2299 | Mouse Moth **31** | 2278 | Poplar Grey | 1904 | Scorched Wing **13** | 1876 | Small Yellow Wave **12** | 648 | White-shoul'd House, micro |
| 2221 | Mullein **28** | 1981 | Poplar Hawk **19** | 1915 | September Thorn | 2198 | Smoky Wainscot | 2203 | White-speck |
| 1689 | Mullein Wave **8** | 1655 | Poplar Lutestring **39** | 1879 | Seraphim **8** | 2477 | Snout **35** | 1835 | White-spotted Pug **11** |
| 2063 | Muslin Moth **23** | 1714 | Portland Ribbon Wave **39** | 2126 | Setaceous Hebrew Ch'ctr **25** | 2264.1 | Southern Chestnut **42** | 1937 | Willow Beauty **16** |
| 1846 | Narrow-winged Pug **11** | 2186 | Powdered Quaker **27** | 1732 | Shaded Broad-bar **9** | 2197 | Southern Wainscot **28** | 428 | Willow Ermine, micro |
| 2132 | Neglected Rustic **25** | 1784 | Pretty Chalk Carpet **40** | 2216 | Shark | 2053 | Speckled Footman **22, front** | 1799 | Winter **10** |
| 1823 | Netted Pug **40** | 1806 | Pretty Pinion **40** | 1794 | Sharp-angled Carpet **10** | 1909 | Speckled Yellow **13** | 1739 | Wood Carpet **39** |
| 385 | Nettle-tap, micro **1** | 1976 | Privet Hawk **17** | 1890 | Sharp-angled Peacock | 2450 | Spectacle **35** | 1830 | Wormwood Pug |
| 2432 | Ni **35** | 1752 | Purple Bar **9** | 2147 | Shears **26** | 1212 | Spotted Shoot, micro **3** | 1968 | Yellow Belle **17** |
| 2104 | Northern Rustic **41** | 2122 | Purple Clay **25** | 2201 | Shore Wainscot **42** | 1932 | Spring Usher **14** | 1659 | Yellow Horned **7** |
| 1800 | Northern Winter **10** | 1919 | Purple Thorn **13** | 2077 | Short-cloaked | 1769 | Spruce Carpet **10** | 1742 | Yellow Shell **9** |
| 1795 | November Moth **10** | 1698 | Purple-bordered Gold **8** | 1746 | Shoulder Stripe | 1949 | Square Spot **16** | 1883 | Yellow-barred Brindle **12** |
| 2145 | Nutmeg | 1995 | Puss Moth **20** | 2205 | Shoulder-striped Wainscot **28** | 2134 | Square-spot Rustic **25** | 2264 | Yellow-line Quaker **29** |
| 2425 | Nut-tree Tussock **34** | 2139 | Red Chestnut **26** | 2092 | Shuttle Shaped Dart **24** | 2474 | Straw Dot **35** | 2030 | Yellow-tail |
| 1930 | Oak Beauty **14** | 1724 | Red Twin-spot Carpet **9** | 2391 | Silky Wainscot **34** | 2303 | Straw Underwing **31** | | |
| 1637 | Oak Eggar **6** | 2452 | Red Underwing **35** | 1704 | Silky Wave **39** | 1864 | Streak **12** | | |
| 1646 | Oak Hook-tip **6** | 1010 | Red-barred Tortrix, micro | 2175 | Silurian **41** | 1747 | Streamer **9** | | |
| 1658 | Oak Lutestring **39** | 1760 | Red-green Carpet **9** | 2181 | Silver Cloud **42** | 1990 | Striped Hawk **41** | | |
| 2423 | Oak Nycteoline **34** | 2263 | Red-line Quaker **29** | 2412 | Silver Hook **34** | 2196 | Striped Wainscot | | |
| 2246.1 | Oak Rustic **43** | 2039 | Red-necked Footman **22** | 2441 | Silver Y **34** | 2268 | Suspected **30** | | |
| 1853 | Oak-tree Pug **11** | 160 | Reed Leopard **37** | 1727 | Silver-ground Carpet | 2007 | Swallow Prominent | | |
| 1719 | Oblique Carpet **8** | 1713 | Riband Wave **8** | 1708 | Single-dotted Wave **8** | 1922 | Swallow-tailed Moth **14** | | |
| 2204 | Obscure Wainscot | 1939 | Ringed Carpet | 169 | Six-spot Burnet | 2279 | Sycamore | | |
| 1844 | Ochreous Pug | 1348 | Ringed China-mark, pyrale **4** | 2133 | Six-striped Rustic **25** | 2339 | Tawny Marbled Minor **32** | | |
| 2300 | Old Lady **31** | 1802 | Rivulet **11** | 227 | Skin Moth, micro | 2235 | Tawny Pinion **28** | | |
| 2312 | Olive | 2037 | Rosy Footman **22** | 2335 | Slender Brindle **32** | 2167 | Tawny Shears | | |
| 2043 | Orange Footman **22** | 2396 | Rosy Marbled | 1811 | Slender Pug | 1893 | Tawny-barred Angle **12** | | |
| 2271 | Orange Sallow | 2342 | Rosy Minor **33** | 2305 | Small Angle Shades **32** | 1838 | Tawny-speckled Pug **11** | | |
| 15 | Orange Swift **1** | 2361 | Rosy Rustic **33** | 2075 | Small Black Arches **41** | 1458 | Thistle Ermine, pyrale **5** | | |
| 1661 | Orange Underwing | 1691 | Rosy Wave **39** | 1690 | Small Blood-vein **8** | 1816 | Toadflax Pug | | |
| 425 | Orchard Ermine, micro | 2035 | Round-winged Muslin **22** | 1925 | Small Brindled Beauty | 1867 | Treble Bar **12** | | |
| 1926 | Pale Brindled Beauty **14** | 2064 | Ruby Tiger | 1354 | Small China-mark, pyrale | 1711 | Treble Brown Spot | | |
| 2389 | Pale Mottled Willow | 1779 | Ruddy Highflyer **10** | 2017 | Small Chocolate-tip **21** | 2380 | Treble Lines **34** | | |
| 1944 | Pale Oak Beauty **16** | 2338 | Rufous Minor | 2331 | Small Clouded Brindle | 2118 | True Lover's Knot | | |
| 2236 | Pale Pinion **28** | 1398 | Rush Veneer, pyrale **4** | 2345 | Small Dotted Buff **33** | 2087 | Turnip **24** | | |
| 2011 | Pale Prominent **21** | 2382 | Rustic | 1707 | Small Dusty Wave | 1288 | Twenty-plume **3** | | |
| 2148 | Pale Shining Brown **41** | 2334 | Rustic Shoulder-knot **43** | 1633 | Small Eggar **37** | 1809 | Twin-spot Carpet **40** | | |
| 2028 | Pale Tussock **22** | 2274 | Sallow **30** | 1992 | Small Elephant Hawk **19** | 2189 | Twin-spotted Quaker **27** | | |
| 2158 | Pale-shouldered Brocade **26** | 377 | Sallow Clearwing | 1673 | Small Emerald **8** | 2370 | Twin-spotted Wainscot **33** | | |

# READING, REFERENCES & WEBSITES

**Field Guide to the Moths of Great Britain & Ireland** *by* **Paul Waring & Martin Townsend**
*Comprehensive information. 1,600 superb illustrations by Richard Lewington and fine larval photographs.*

**Moths of the British Isles** *by* **Bernard Skinner**.
*Standard text book covering all known resident and migrant macromoths. Shows 'set' specimens, all life size.*

**British Pyralid Moths** *by* **Barry Goater**.
*Similar to above, but for pyrales. (A group of the larger micros)*

**Caterpillars of the British Isles** *by* **Jim Porter**.
*Photographs of larvae of all known British macromoths and butterflies.*

### www.ukmoths.org.uk
*Brilliant site run by Ian Kimber aiming to show photographs of all live UK moths, currently 1618 species ex approx 2400.*

### www.portlandbirdobs.btinternet.co.uk
*Martin Cade runs moth lights on Portland most nights and posts stunning photographs of migrant moths and birds most days.*

### www.migrantmoth.com
*Steve Nash dedicates his spare time to keeping track of movements of migrant moth species.*

### www.dorsetmothgroup.org.uk
*A must for 'mothers' in Dorset. Site run by Terry Box & Jim Fradgley. Newsletter editor Gordon Hopkins*

### www.butterfly-conservation.org  &  www.dorsetbutterflies.com
*Butterfly Conservation, Manor Yard, East Lulworth, Wareham, Dorset BH20 5QP  tel 0870 774 4309*
*Leads action on UK BAP species. Producers of the 2006 report "The state of Britain's larger moths" jointly with Rothamsted Research*

### www.dorsetwildlife.co.uk
*Dorset Wildlife Trust, Brooklands Farm, Forston, Dorchester DT2 7AA  tel 01305 264620*

### www.atropos.info  &  www.nationalmothnight.info
*Editor: Mark Tunmore, 36 Tinker Lane, Meltham, Holmfirth, West Yorkshire HD9 4EX*
*Atropos is the only independent journal devoted to the Lepidoptera and Odonata of the British Isles. Well written and illustrated. Moth book stockists.*

### www.amentsoc.org
*Amateur Entomologists' Society, PO Box 8774, London SW7 5ZG*

### www.benhs.org.uk
*British Entomological & Natural History Society, Dinton Pastures Country Park, Davis Street, Hurst, Reading, Berks RG10 0TH*

### www.watdon.com
*Watkins & Doncaster, PO Box 5, Cranbrook, Kent TN18 5EZ.  Entomological equipment, books etc. Established 1874, founder of London Zoo insect house.*

### www.angleps.btinternet.co.uk
*Anglian Lepidopterist Supplies, Station Road, Hindolveston, Norfolk NR20 5DE. Entomology supplies, very informative site.*

*****